SpringerBriefs in Applied Sciences and Technology

Multiphase Flow

W0230301

Series editors

Lixin Cheng, Portsmouth, UK
Dieter Mewes, Hannover, Germany

For further volumes:
http://www.springer.com/series/11897

Zhong-Ke Gao · Ning-De Jin
Wen-Xu Wang

Nonlinear Analysis
of Gas-Water/Oil-Water
Two-Phase Flow
in Complex Networks

 Springer

Zhong-Ke Gao
Ning-De Jin
School of Electrical Engineering
    and Automation
Tianjin University
Tianjin
People's Republic of China

Wen-Xu Wang
School of Electrical, Computer and Energy
    Engineering
Arizona State University
Tempe
USA

ISSN 2191-530X   ISSN 2191-5318   (electronic)
ISBN 978-3-642-38372-4   ISBN 978-3-642-38373-1   (eBook)
DOI 10.1007/978-3-642-38373-1
Springer Heidelberg New York Dordrecht London

Library of Congress Control Number: 2013955903

Printed on acid-free paper

Springer is part of Springer Science+Business Media (www.springer.com)

Acknowledgments

This work was supported by National Natural Science Foundation of China (Grant Nos. 41174109, 50974095) and National Science and Technology Major Projects (Grant No. 2011ZX05020-006). Z. K. Gao was also supported by National Natural Science Foundation of China under Grant No. 61104148, Specialized Research Fund for the Doctoral Program of Higher Education of China under Grant No. 20110032120088, and Elite Scholar Program of Tianjin University.

Contents

Nomenclature

A	Adjacency matrix
a_1, a_2, a_3	First non-trivial eigenvectors
b_i	Betweenness of a node i
c_k	Frequency characteristic quantities
C_{ij}	Correlation coefficient
$C^L(v)$	Local clustering coefficient
C_i	Clustering coefficient of node i
$<C>$	Average of weighted clustering coefficient for all nodes
C	Correlation matrix
D	Diagonal matrix
d_{ij}	Phase-space distance between two vector points
d_l	System parameters of linear prediction model
e_{ij}	Fraction of edges in the network that connect nodes in community i to those in community j
E	Network information entropy
E_N	Normalized network information entropy
G	System gain of linear prediction model
H	Standard matrix
I	Shannon entropy
I_t	Input signals for linear prediction model
ID	Pipe inner diameter, mm
K	Diagonal matrix with the ith element equal to the degree of ith node
k	Degree (or connectivity)
k_B	Boltzmann constant
$k_{nn}(k)$	Average degree of the nearest neighbors
L	Length of segment of time series
M	Dimension of the characteristic vector
m	Embedding dimension of phase space
n	Number of nodes contained in the network

n_{jk}	Number of shortest paths connecting node j and k
$n_{jk}(i)$	Number of shortest paths connecting j and k and passing through i
$p(k)$	Probability that a node chosen uniformly at random has degree k
$P(i)$	Importance of node i
$p(k'|k)$	Conditional probability that an edge of degree k connects a node with degree k'
Q	Modularity
Q_g	Gas flowrate, m³/h
Q_o	Oil flowrate, m³/h
Q_w	Water flowrate, m³/h
R_{ij}	Recurrence matrix
r_c	Threshold
r	Pearson correlation coefficient
$SampEn$	Sample entropy
$\mathbf{S_w}$	wth segment of a time series
t	Time, s
$\mathbf{T_i}, \mathbf{T_j}$	Characteristic vector
U_{t-l}	Unknown input signals for linear prediction model
U_{sg}	Gas superficial velocity, m/s
U_{so}	Oil superficial velocity, m/s
U_{sw}	Water superficial velocity, m/s
$\vec{X}_i, \vec{X}_j$	Vector points of phase space
$x(t)$	Conductance fluctuating signals measured from VMEA sensor, Volt
x_{max}	Maximum value of the signals measured from VMEA sensor, Volt
x_{min}	Minimum value of the signals measured from VMEA sensor, Volt
$\bar{x}$	Mean value of the signals measured from VMEA sensor, Volt
x_{sta}	Standard deviation of the signals measured from VMEA sensor
x_{dis}	Dissymmetry coefficient of the signals measured from VMEA sensor
x_{kur}	Kurtosis coefficient of the signals measured from VMEA sensor
z_i	ith point of a time series

Greek letters

μ	Lagrange multiplier
γ	Power-law exponent of degree distribution
τ	Time delay
σ_q^2	Variance of q_k and $\sigma_q^2 = \sum_k k^2 q_k - \left(\sum_k k q_k\right)^2$

E	Threshold
ε_i	Gaussian iid random variable with zero mean and a standard deviation of 1
η	Strength of noise

Subscripts

g	Gas
o	Oil
w	Water

Abstract

Understanding the dynamics of multiphase flows has been a challenge in the fields of nonlinear dynamics and fluid mechanics. This book reviews our works on two-phase flow dynamics in combination with complex network theory. We systematically carried out gas-water/oil-water two-phase flow experiments for measuring the time series of flow signals, which is studied in terms of the mapping from time series to complex networks. Three network mapping methods were proposed for the analysis and identification of flow patterns, i.e., flow pattern complex network (FPCN), fluid dynamic complex network (FDCN), and fluid structure complex network (FSCN). Through detecting the community structure of FPCN based on K-means clustering, distinct flow patterns can be successfully distinguished and identified. A number of FDCNs under different flow conditions were constructed in order to reveal the dynamical characteristics of two-phase flows. The FDCNs exhibit universal power-law degree distributions. The power-law exponent and the network information entropy are sensitive to the transition among different flow patterns, which can be used to characterize nonlinear dynamics of the two-phase flow. FSCNs were constructed in the phase space through a general approach that we introduced. The statistical properties of FSCN can provide quantitative insight into the fluid structure of two-phase flow. In addition, the directed weighted complex network, Markov transition probability-based network, recurrence network also have been employed to investigate the experimental two-phase flow and many fascinating results are demonstrated in detail. These interesting and significant findings suggest that complex networks can be a potentially powerful tool for uncovering the nonlinear dynamics of two-phase flows.

Keywords Multiphase flow · Complex network · Time series analysis · Nonlinear dynamics · Sensors

Chapter 1
Introduction

Gas-water/oil-water two-phase flow is quite common in lubrication, spray processes, nuclear reactor cooling and well bores. The behaviors of two-phase flow under a wide range of flow conditions and inclination angles constitute an outstanding interdisciplinary problem with significant applications to the petroleum industry. Understanding the dynamics of flow patterns is a crucial issue. Due to the interplay among many complex factors such as fluid turbulence, phase interfacial interaction, and local relative movements between phases, two-phase flow exhibits highly irregular, random, and unsteady flow structure.

Earlier investigations of two-phase flow were mainly focused on experimental observations. For example, Hewitt [1] observed bubbly flow, slug flow, churn flow, annular flow and wispy-annular flow and Taitel et al. [2] observed bubbly flow, slug flow, churn flow, annular flow and mist flow in the study of vertical gas-water two-phase flow. Hill and Oolman [3] observed, in a 152 mm inner-diameter (ID) pipe, a kind of segregated flow patterns where the water phase exists in most of the pipe but the flow tends to reverse near the bottom of the pipe. They observed that a small change in the deviation angle can cause a large change in the velocity profile distribution. Vigneaux et al. [4] measured the inclined oil-water flows in a 200 mm ID pipe by using a high-frequency impedance probe and observed the occurrence of two main flow patterns: dispersed oil in water-pseudoslugs (PS) flow and in water-countercurrent (CT) flow. Flores et al. [5] conducted a comprehensive experimental study of vertical and inclined oil-water flows with a 50.8 mm ID pipe, and classified seven flow patterns in inclined oil-water flows with four water-dominated, two oil-dominated and a transitional flow (TF) pattern.

In the 1990s, numerical simulation methods began to be widely used in the study of two-phase flows. Take the inclined oil-water flow as an example. Mobbs and Lucas [6] proposed, for inclined liquid-liquid flow, a large-amplitude turbulence model that incorporates qualitatively Kelvin-Helmholtz eddy characteristics. They found that the eddies grow and collapse periodically and their amplitudes can reach the value of the pipe diameter. Lucas [7] proposed a mathematical model of velocity profile for inclined oil-water flows, where predictions of local velocities agree with experiments but only for the upper and central part of the pipe. Lucas and Jin [8, 9] studied a drift-velocity model and homogeneous velocity

Z.-K. Gao et al., *Nonlinear Analysis of Gas-Water/Oil-Water
Two-Phase Flow in Complex Networks*, SpringerBriefs on Multiphase Flow,
DOI: 10.1007/978-3-642-38373-1_1, © The Author(s) 2014

measurement in inclined oil-water pipe flows, and demonstrated that the phase-distribution parameter and single-droplet terminal rise velocity can be greatly affected by inclination angles.

Quite recently, due to the development of signal processing techniques, there has been much progress in the software measurement techniques. Mi et al. [10] applied a neural network to two-phase flow pattern identification in a vertical channel using signals from electric capacitance probes. Warsito and Fan [11] exploited a neural multi-criterion optimization image reconstruction technique based on network for imaging multiphase flow systems from electrical capacitance tomography (ECT). Yan et al. [12] identified the two-component flow regimes by means of back-propagation networks. Daw et al. [13–15] interpreted experimental pressure-drop measurements from a complex gas-solids flow system in terms of chaotic time-series analysis, and discussed issues of reconstructing attractors from experimental chaotic time-series data using Taken's method of delays. Jin et al. [16, 17] established a general description method of chaotic attractor morphological characteristic using referenced sections, and proposed a new method for flow pattern classification by combining the chaotic attractor morphological feature parameters of different dimensions.

Despite these contributions, there still exist significant challenges in the study of two-phase flow. For example, detecting transitional flow is an unsolved problem, yet. Due to the complexity of the problem, analytical approaches are usually infeasible. The approach of nonlinear time series analysis [18] also has severe limitations as applied to oil-water two-phase flows, mainly due to the occurrence of CT flow pattern triggered by the gravitational component normal to the flow direction. In particular, in Ref. [16] we pointed out that, although dispersion Oil-in-water PS and CT flows can be distinguished by the methods of recurrence plot [19, 20] and attractor-geometry morphological mapping [21], these methods appear to be ineffective for transitional flows. So far there has been no satisfactory understanding of the underlying dynamics which control the evolution of patterns in such flows.

Meanwhile, the past few years have witnessed dramatic advances in the field of complex networks since the publication of the seminal works of Watts and Strogatz [22] as well as Barabási and Albert [23]. Complex networks, which have been observed to arise naturally in a vast range of physical phenomena, can describe any complex system that contains massive units (or subsystems) with nodes representing the component units and edges standing for the interactions between them. A lot of complex systems have been examined in the viewpoint of complex networks. Examples include World Wide Web [24], metabolic networks [25], protein networks in the cell [26], traffic networks [27], scientific collaboration networks [28], earthquake networks [29] and functional modularity networks [30]. These empirical studies have inspired researchers to develop a variety of techniques and models to help us understand or predict the behavior of complex systems [31–37]. Recently, complex network theory has been introduced into the study of time series analysis and complex network analysis of time series has attracted much attention from various research fields on account of its significant importance [38–59]. Complex networks from time series can provide important complementary properties of dynamic system based on spatial dependences between individual observations, which cannot be captured by the existing time-series analysis methods. Different approaches for mapping a time series into a complex network have been proposed, such as the approaches based on the concept of quasi-periodic cycle [38, 39], correlations [40], visibility [41–44], transition probabilities [45–47], and recurrence analysis (phase-space reconstruction) [48–56], etc. Complex network analysis of time series has been successfully applied in many research fields, such as network topology estimation [57–59], climate system [60–62], financial system [63–65], human gait [66], human brain [30], human ventricular fibrillation [67], gene system [68], aperture evolution in a rock joint [69] and friction networks in nucleation processes [70, 71] etc. Complex networks, which provide us with a new viewpoint and an effective tool for understanding a complex system from the relations between the elements in a global way, not only may be a powerful tool for revealing information embedded in time series but also can be used for studying nonlinear dynamic systems that can not be perfectly described by theoretical model.

Quite recently, the approach of complex networks has been introduced into the study of two-phase flows [72–78]. Bridging time series analysis and complex networks can be an appealing approach for experimental data analysis and pattern recognition. Based on experimental measured time series from a two-phase flow, artificial complex networks can be constructed. In this book, gas-water and oil-water two-phase flow will be investigated from the point of view of complex networks.

References

1. G.F. Hewitt, *Measurement of Two-phase Flow Parameters* (Academic Press, London, 1978)
2. Y. Taitel, D. Barnea, Modelling flow pattern transitions for steady upward gas-water flow in vertical tubes. AIChE J. **26**(3), 345–354 (1980)
3. A.D. Hill, T. Oolman, Production logging tool behavior in two-phase inclined flow. J. Petrol. Technol. **34**, 2432–2440 (1982)
4. P. Vigneaux, P. Chenais, J.P. Hulin, Liquid-liquid flows in an inclined pipes. AIChE J. **34**(5), 781–789 (1988)
5. J.G. Flores, X.T. Chen, C. Sarica, J.P. Brill, Characterization of oil-water flow patterns in vertical and deviated wells. SPE Prod. Facil. **14**, 102–109 (1999)
6. S.D. Mobbs, G.P. Lucas, A turbulence model for inclined, bubbly flow. Appl. Sci. Res. **51**, 263–268 (1993)
7. G.P. Lucas, Modelling velocity profiles in inclined multiphase flow to provide a priori information for flow imaging. Chem. Eng. J. **56**, 167–173 (1995)
8. G.P. Lucas, N.D. Jin, Investigation of a drift velocity model for predicting superficial velocities of oil and water in inclined oil-in-water pipe flows with a centre body. Meas. Sci. Technol. **12**, 1546–1554 (2001)
9. G.P. Lucas, N.D. Jin, Measurement of the homogeneous velocity of inclined oil-in-water flows using a resistance cross correlation flow meter. Meas. Sci. Technol. **12**, 1529–1537 (2001)

10. Y. Mi, M. Ishii, L.H. Tsoukalas, Flow regime identification methodology with neural networks and two-phase flow model. Nucl. Eng. Des. **204**(1–3), 87–100 (2001)
11. W. Warsito, L.S. Fan, Neural network based multi-criterion optimization image reconstruction technique for imaging two- and three-phase flow systems using electrical capacitance tomography. Meas. Sci. Technol. **12**(12), 2198–2210 (2001)
12. H. Yan, Y.H. Liu, C.T. Liu, Identification of flow regimes using back-propagation networks trained on simulated data based on a capacitance tomography sensor. Meas. Sci. Technol. **15**(2), 432–436 (2004)
13. C.S. Daw, W.F. Lawkins, D.J. Downing, N.E. Clapp Jr, Chaotic characteristics of a complex gas-solids flow. Phys. Rev. A **41**(2), 1179–1181 (1990)
14. W.F. Lawkins, C.S. Daw, D.J. Downing, N.E. Clapp Jr, Role of low-pass filtering in the process of attractor reconstruction from experimental chaotic time series. Phys. Rev. E. **47**(4), 2520–2535 (1993)
15. C.S. Daw, C.E.A. Finney, M. Vasudevan, N.A. vanGoor, K. Nguyen, D.D. Bruns, E.J. Kostelich, C. Grebogi, E. Ott, J.A. Yorke, Self-organization and chaos in a fluidized bed. Phys. Rev. Lett. **75**(12), 2308–2311 (1995)
16. Y.B. Zong, N.D. Jin, Z.Y. Wang, Z.K. Gao, C. Wang, Nonlinear dynamic analysis of large diameter inclined oil-water two phase flow pattern. Int. J. Multiph. Flow **36**(3), 166–183 (2010)
17. Z.Y. Wang, N.D. Jin, Z.K. Gao, Y.B. Zong, T. Wang, Nonlinear dynamical analysis of large diameter vertical upward oil-gas-water three-phase flow pattern characteristics. Chem. Eng. Sci. **65**(18), 5226–5236 (2010)
18. H. Kantz, T. Schreiber, *Nonlinear Time Series Analysis* (Cambridge University Press, Cambridge, 1997)
19. J.P. Eckmann, S.O. Kamphorst, D. Ruelle, Recurrence plots of dynamical systems. Europhys. Lett. **5**, 973–977 (1987)
20. N. Marwan, N. Wessel, U. Meyerfeldt, A. Schirdewan, J. Kurths, Recurrence plot based measures of complexity and its application to heart rate variability data. Phys. Rev. E **66**, 26702 (2002)
21. M. Annunziato, H.D.I. Abarbanel, in *Proceedings of International Conference on Soft Computing*, Genova, Italy, (1999)
22. D.J. Watts, S.H. Strogatz, Collective dynamics of 'small-world' networks. Nature **393**(6684), 440–442 (1998)
23. A.-L. Barabási, R. Albert, Emergence of scaling in random networks. Science **286**(5439), 509–512 (1999)
24. R. Albert, H. Jeong, A.-L. Barabási, Diameter of the world-wide web. Nature **401**, 130–131 (1999)
25. H. Jeong, B. Tombor, R. Albert, Z.N. Oltvai, A.-L. Barabási, The large-scale organization of metabolic networks. Nature **407**(6804), 651–654 (2000)
26. H. Jeong, S. Mason, A.-L. Barabási, Z.N. Oltvai, Lethality and centrality in protein networks. Nature **411**(6683), 41–42 (2001)
27. X.G. Li, Z.Y. Gao, K.P. Li, X.M. Zhao, Relationship between microscopic dynamics in traffic flow and complexity in networks. Phys. Rev. E **76**(1), 016110 (2007)
28. M.E.J. Newman, The structure of scientific collaboration networks. Proc. Natl. Acad. Sci. U.S.A. **98**, 404–409 (2001)
29. A. Sumiyoshi, S. Norikazu, Complex earthquake networks: hierarchical organization and assortative mixing. Phys. Rev. E **74**, 026113 (2006)
30. M. Chavez, M. Valencia, V. Navarro, V. Latora, J. Martinerie, Functional modularity of background activities in normal and epileptic brain networks. Phys. Rev. Lett. **104**(11), 118701 (2010)
31. R. Albert, A.L. Barabási, Statistical mechanics of complex networks. Rev. Mod. Phys. **74**, 47–97 (2002)
32. S. Boccaletti, V. Latora, Y. Moreno, M. Chavez, D.U. Hwang, Complex networks: structure and dynamics. Phys. Rep. **424**, 175–308 (2006)

33. W.X. Wang, B.H. Wang, B. Hu, G. Yan, Q. Ou, General dynamics of topology and traffic on weighted technological networks. Phys. Rev. Lett. **94**, 188702 (2005)
34. L. Huang, K. Park, Y.C. Lai, L. Yang, K.Q. Yang, Abnormal synchronization in complex clustered networks. Phys. Rev. Lett. **97**, 164101 (2006)
35. Z.Q. Jiang, W.X. Zhou, B. Xu, W.K. Yuan, Process flow diagram of an ammonia plant as a complex network. AIChE J. **53**, 423–428 (2007)
36. W.X. Wang, L. Huang, Y.C. Lai, Universal dynamics on complex networks. Europhys. Lett. **87**, 18006 (2009)
37. J. Ren, W.X. Wang, B. Li, Y.C. Lai, Noise bridges dynamical correlation and topology in coupled oscillator networks. Phys. Rev. Lett. **104**, 058701 (2010)
38. J. Zhang, M. Small, Complex network from pseudoperiodic time series: topology versus dynamics. Phys. Rev. Lett. **96**, 238701 (2006)
39. J. Zhang, J. Sun, X. Luo, K. Zhang, T. Nakamura, M. Small, Characterizing pseudoperiodic time series through complex network approach. Physica D **237**, 2856–2865 (2008)
40. Y. Yang, H.J. Yang, Complex network-based time series analysis. Physica A **387**, 1381–1386 (2008)
41. L. Lacasa, B. Luque, F. Ballesteros, J. Luque, J.C. Nuno, From time series to complex networks: the visibility graph. Proc. Natl. Acad. Sci. U.S.A. **105**, 4972–4975 (2008)
42. L. Lacasa, R. Toral, Description of stochastic and chaotic series using visibility graphs. Phys. Rev. E **82**, 036120 (2010)
43. B. Luque, L. Lacasa, F.J. Ballesteros, A. Robledo, Feigenbaum graphs: a complex network perspective of chaos. PLoS ONE **6**, e22411 (2011)
44. C. Liu, W.X. Zhou, W.K. Yuan, Statistical properties of visibility graph of energy dissipation rates in three-dimensional fully developed turbulence. Physica A **389**, 2675–2681 (2010)
45. P. Li, B.H. Wang, Extracting hidden fluctuation patterns of Hang Seng stock index from network topologies. Physica A **378**, 519–526 (2007)
46. A.H. Shirazi, G.R. Jafari, J. Davoudi, J. Peinke, M.R.R. Tabar, M. Sahimi, Mapping stochastic processes onto complex networks. J. Stat. Mech.-Theo. Exp. P07046 (2009)
47. A.S.L.O. Campanharo, M.I. Sirer, R.D. Malmgren, F.M. Ramos, L.A.N. Amaral, Duality between time series and networks. PLoS ONE **6**, e23378 (2011)
48. X. Xu, J. Zhang, M. Small, Superfamily phenomena and motifs of networks induced from time series. Proc. Natl. Acad. Sci. U.S.A. **105**, 19601–19605 (2008)
49. N. Marwan, J.F. Donges, Y. Zou, R.V. Donner, J. Kurths, Complex network approach for recurrence analysis of time series. Phys. Lett. A **373**, 4246–4254 (2009)
50. Z.K. Gao, N.D. Jin, Complex network from time series based on phase space reconstruction. Chaos **19**, 033137 (2009)
51. Z.K. Gao, N.D. Jin, A directed weighted complex network for characterizing chaotic dynamics from time series. Nonlinear Anal. Real World Appl. **13**, 947–952 (2012)
52. R.V. Donner, Y. Zou, J.F. Donges, N. Marwan, J. Kurths, Recurrence networks-a novel paradigm for nonlinear time series analysis. New J. Phys. **12**, 033025 (2010)
53. R.V. Donner, J. Heitzig, J.F. Donges, Y. Zou, N. Marwan, J. Kurths, The geometry of chaotic dynamics—a complex network perspective. Eur. Phys. J. B. **84**, 653–672 (2011)
54. R.V. Donner, M. Small, J.F. Donges, N. Marwan, Y. Zou, R. Xiang, J. Kurths, Recurrence-based time series analysis by means of complex network methods. Int. J. Bifurcat. Chaos **21**(4), 1019–1046 (2011)
55. J.F. Donges, J. Heitzig, R.V. Donner, J. Kurths, Analytical framework for recurrence network analysis of time series. Phys. Rev. E **85**, 046105 (2012)
56. R. Xiang, J. Zhang, X.K. Xu, M. Small, Multiscale characterization on recurrence-based phase space networks constructed from time series. Chaos **22**, 013107 (2012)
57. W.X. Wang, R. Yang, Y.C. Lai, V. Kovanis, C. Grebogi, Predicting catastrophes in nonlinear dynamical systems by compressive sensing. Phys. Rev. Lett. **106**, 154101 (2011)
58. W.X. Wang, R. Yang, Y.C. Lai, V. Kovanis, M.A.F. Harrison, Time-series-based prediction of complex oscillator networks via compressive sensing. Europhys. Lett. **94**, 48006 (2011)

59. R.Q. Su, W.X. Wang, Y.C. Lai, Detecting hidden nodes in complex networks from time series. Phys. Rev. E. **85**, 065201(R) (2012)
60. J.F. Donges, Y. Zou, N. Marwan, J. Kurths, The backbone of the climate network. Europhys. Lett. **87**, 48007 (2009)
61. O. Guez, A. Gozolchiani, Y. Berezin, S. Brenner, S. Havlin, Climate network structure evolves with North Atlantic Oscillation phases. Europhys. Lett. **98**, 38006 (2012)
62. J.F. Donges, R.V. Donner, M.H. Trauth, N. Marwan, H.J. Schellnhuber, J. Kurths, Nonlinear detection of paleoclimate-variability transitions possibly related to human evolution. Proc. Natl. Acad. Sci. U.S.A. **108**, 20422 (2011)
63. D.M. Song, Z.Q. Jiang, W.X. Zhou, Statistical properties of world investment networks. Physica A **388**, 2450–2460 (2009)
64. R.V. Donner, J.F. Donges, Y. Zou, N. Marwan, J. Kurths, Recurrence-based evolving networks for time series analysis of complex systems, in *proceedings of the International Symposium on Nonlinear Theory and its Applications (NOLTA2010)*, Krakow(6165), pp. 87–90
65. M.C. Qian, Z.Q. Jiang, W.X. Zhou, Universal and nonuniversal allometric scaling behaviors in the visibility graphs of world stock market indices. J. Phys. A: Math. Theor. **43**, 335002 (2010)
66. J. Zhang, K. Zhang, J.F. Feng, M. Small, Rhythmic dynamics and synchronization via dimensionality reduction: application to human gait. PLoS Comput. Biol. **6**, e1001033 (2010)
67. X. Li, Z. Dong, Detection and prediction of the onset of human ventricular fibrillation: an approach based on complex network theory. Phys. Rev. E. **84**, 062901 (2011)
68. S. Hempel, A. Koseska, J. Kurths, Z. Nikoloski, Inner composition alignment for inferring directed networks from short time series. Phys. Rev. Lett. **107**, 054101 (2011)
69. H.O. Ghaffari, M. Sharifzadeh, M. Fall, Analysis of aperture evolution in a rock joint using a complex network approach. Int. J. Rock Mech. Min. Sci. **47**, 17–29 (2010)
70. H.O. Ghaffari, R.P. Young, Topological complexity of frictional interfaces: friction networks. Nonlinear Processes Geophys. **19**, 215–225 (2012)
71. H.O. Ghaffari, R.P. Young, Network configurations of dynamic friction patterns. Europhys. Lett. **98**, 48003 (2012)
72. Z.K. Gao, N.D. Jin, Flow-pattern identification and nonlinear dynamics of gas-water two-phase flow in complex networks. Phys. Rev. E **79**(6), 066303 (2009)
73. Z.K. Gao, N.D. Jin, W.X. Wang, Y.C. Lai, Motif distributions in phase-space networks for characterizing experimental two-phase flow patterns with chaotic features. Phys. Rev. E **82**(2), 016210 (2010)
74. Z.K. Gao, N.D. Jin, Nonlinear characterization of oil-gas-water three-phase flow in complex networks. Chem. Eng. Sci. **66**, 2660–2671 (2011)
75. Z.K. Gao, N.D. Jin, Characterization of chaotic dynamic behavior in the gas-liquid slug flow using directed weighted complex network analysis. Physica A **391**, 3005–3016 (2012)
76. Z.K. Gao, X.W. Zhang, M. Du, N.D. Jin, Recurrence network analysis of experimental signals from bubbly oil-in-water flows. Phys. Lett. A. **377**, 457–462 (2013)
77. Z.K. Gao, L.D. Hu, N.D. Jin, Markov transition probability-based network from time series for characterizing experimental two-phase flow. Chin. Phys. B. **22**(5), 050507 (2013)
78. Z.K. Gao, X.W. Zhang, N.D. Jin, R.V. Donner, N. Marwan, J. Kurths, Recurrence network from multivariate signals for uncovering dynamic behavior of horizontal oil-water stratified flows, Europhys. Lett. **103**, 50004 (2013)

Chapter 2
Definition of Flow Patterns

2.1 Vertical Upward Gas-Water Flow Patterns

The vertical upward gas-water two-phase flow patterns in a pipe of inner diameter 125 mm can be categorized into five classes on the basis of the visual and video observations and still photography. According to Hewitt [1], the five flow patterns, observed in our experiment, can be defined as follows (Fig. 2.1).

Bubble flow (Fig. 2.1a): This flow pattern occurs at low gas flow rates where the gas phase is approximately uniformly distributed in the form of discrete bubbles in a continuum of liquid phase.

Bubble-slug transitional flow (Fig. 2.1b): This flow pattern is characterized by the non-uniform distribution of the concentration of small bubbles in the flow direction. Small bubble coalescence occasionally arises in the part of high bubble concentration, and as a result, a spherically capped bubble is formed.

Slug flow (Fig. 2.1c): Most of the gas appears in large bullet shaped bubbles, also known as Taylor bubbles, whose diameters almost equal the pipe diameter. The liquid slug area between two Taylor bubbles is filled with small bubbles that are quite similar to those in bubble flow.

Slug-churn transitional flow (Fig. 2.1d): As the gas flow rate increases, for example, the gas-water interface of the larger gas bubble becomes distorted near the nose, but still comparatively smooth in the bottom part of a cylindrical gas bubble.

Churn flow (Fig. 2.1e): Because of instabilities in the slugs, churn flow is a highly disordered flow that arises at high gas flow rates. Churn flow can be interpreted as an irregular, chaotic and disordered slug flow. It is also characterized by an oscillatory flow, associated with the moving of liquid phase alternately upward and downward in the channel.

Z.-K. Gao et al., *Nonlinear Analysis of Gas-Water/Oil-Water*   7
Two-Phase Flow in Complex Networks, SpringerBriefs on Multiphase Flow,
DOI: 10.1007/978-3-642-38373-1_2, © The Author(s) 2014

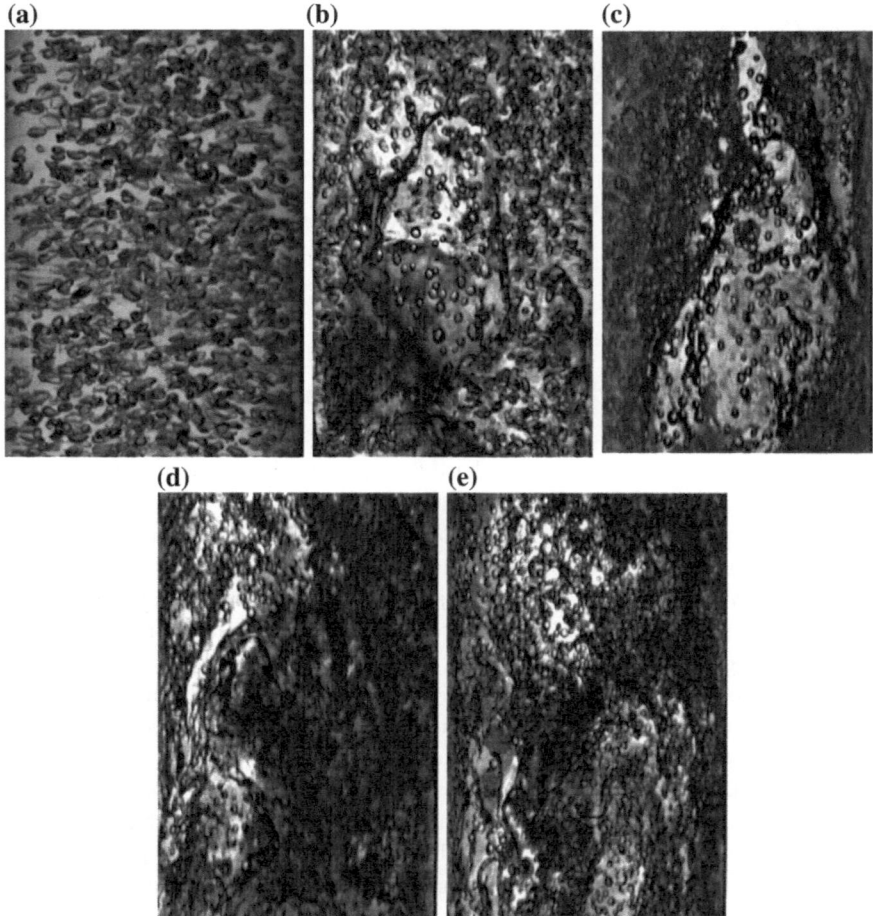

Fig. 2.1 The five vertical upward gas-water two-phase flow patterns recorded by high speed VCR system. **a** Bubble flow; **b** Bubble-slug transitional flow; **c** Slug flow; **d** Slug-churn transitional flow; **e** Churn flow

2.2 Horizontal Gas-Water Flow Patterns

Horizontal gas-liquid stratified flow occurs at very low flow rate. The flow structure of stratified flow is relative stable, the upper part of the pipe is gas phase and the bottom part of the pipe is the water phase. Due to the influence of gravity, there exists a smooth interface between gas and water phase, as shown in Fig. 2.2a–b.

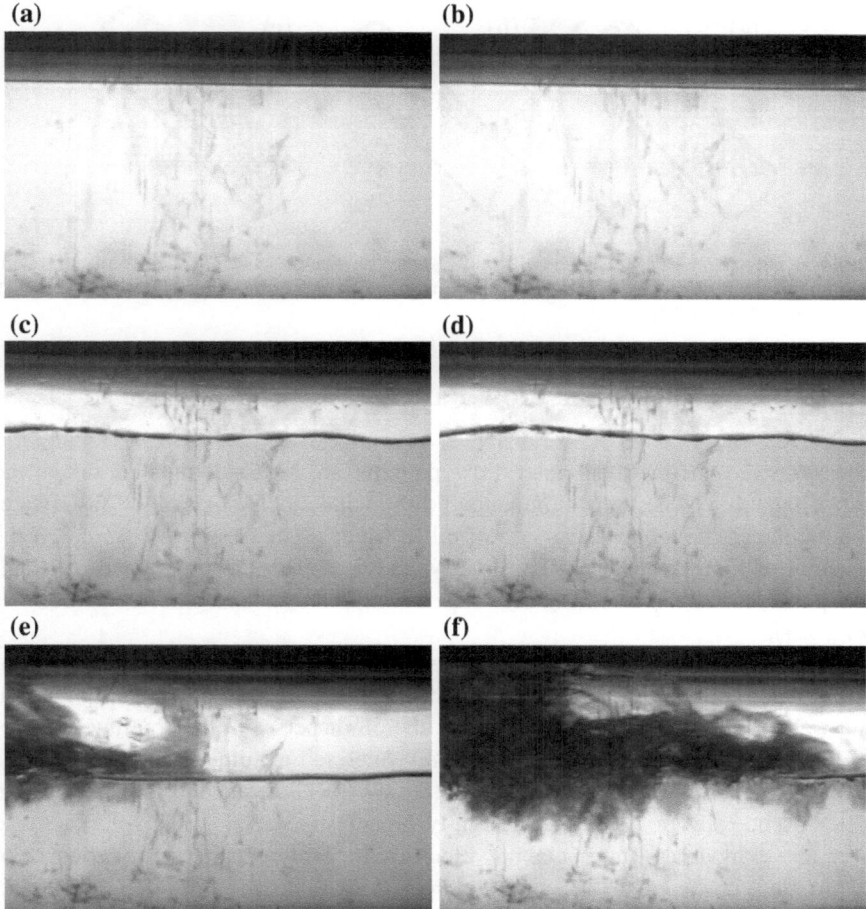

Fig. 2.2 Snapshots for three typical horizontal gas-liquid flow patterns: **a–b** Stratified flow; **c–d** Stratified wavy flow; **e–f** Slug flow

With the increase of gas flow rate, the interface becomes unstable and stratified wavy flow gradually appears with the phenomenon of interface fluctuations, as shown in Fig. 2.2c–d. In this flow pattern transition, due to the increase of flow rate, the turbulence energy will increase. Consequently, the flow structure of stratified wavy flow become unstable gradually and the fluctuations appear in the interface between gas phase and water phase, as shown in Fig. 2.2c–d.

When the gas superficial velocity is high, horizontal slug flow appears. Because of the influence of the turbulence effect, the flow structure of slug flow is very unstable, and compared with that of stratified wavy flow, the interface fluctuations of slug flow become strengthened, as shown in Fig. 2.2e–f.

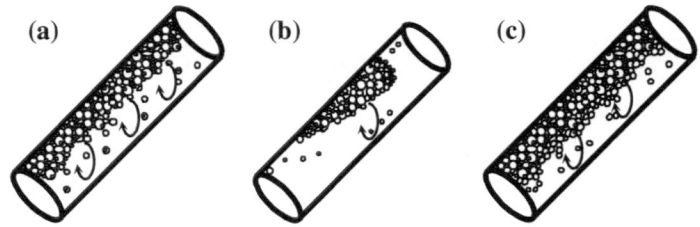

Fig. 2.3 Three water-dominated inclined oil-water flow patterns: **a** Dispersion oil in water-countercurrent flow (D O/W CT); **b** Dispersion oil in water-pseudoslugs flow (D O/W PS); **c** Transitional flow (TF)

2.3 Inclined Oil-Water Flow Patterns

The observed inclined oil-water flow patterns are water-dominated flows and transitional flow. The water-dominated flows include the dispersion of oil in water countercurrent flow and dispersion of oil in water pseudoslug flow. Based on the research of Flores et al. [2], the three typical flow patterns of inclined oil-water two-phase flow, observed in our experiment, can be categorized as follows (Fig. 2.3).

Dispersion Oil in Water-Countercurrent (D O/W CT) flow: The dispersion of oil in water countercurrent flow occurs between low and moderate superficial oil and water velocities. In this flow pattern, the oil disperses in the continuous water as discrete, well rounded, droplets of mostly small to mediums size (see Fig. 2.3a).

Dispersion Oil in Water-Pseudoslugs (D O/W PS) flow: The dispersion of oil in water pseudoslug flow occurs at slightly higher superficial water velocities and from low to moderate superficial oil velocities. In this flow pattern, the sequence of oil droplets observed in the countercurrent flow pattern is interrupted by water breaks, associated with the aggregation and packing of oil droplets at the top of the pipe (see Fig 2.3b).

Transitional flow (TF): The transitional flow occurs at moderate superficial oil velocities and from low to moderate superficial water velocities, in which oil stays at the topside region of the pipe, while at the bottom of the pipe, exist just water containing a few recirculating oil droplets; in the middle region, oil phase and water phase alternately switch (see Fig. 2.3c).

2.4 Vertical Upward Oil-in-Water Flow Patterns

Vertical upward oil-in-water slug flow occurs at low oil-water mixture flowrate, where the small oil bubbles in water continuous phase coalesce to form oil slugs in different sizes. With the coalescence of small oil bubbles becomes more and more, a large number of oil slugs appear.

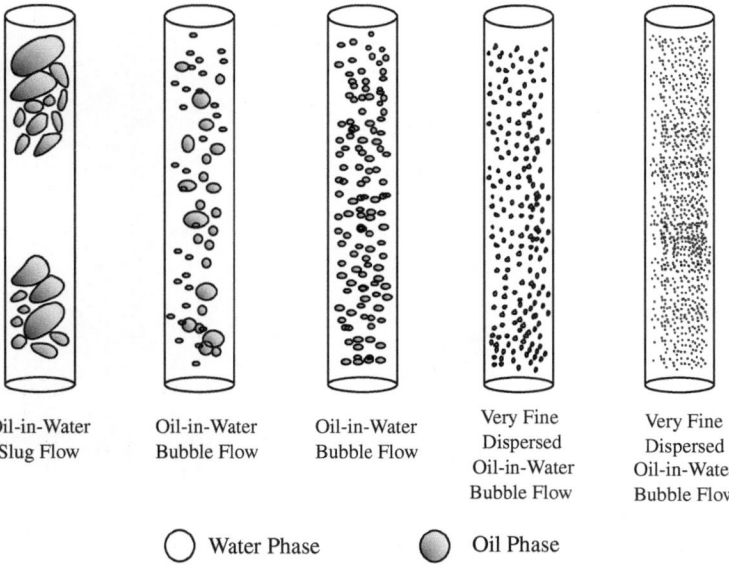

| Oil-in-Water Slug Flow | Oil-in-Water Bubble Flow | Oil-in-Water Bubble Flow | Very Fine Dispersed Oil-in-Water Bubble Flow | Very Fine Dispersed Oil-in-Water Bubble Flow |

◯ Water Phase   ◉ Oil Phase

Fig. 2.4 Three types of vertical upward oil-in-water flow patterns in a small diameter pipe. The size of the oil bubble in the five flow conditions from left to right is approximate 10, 5, 2, 1, 0.5 mm, respectively

With the increase of total velocity of the oil-water mixture flow, the turbulent kinetic energy is strengthened and the oil slugs are dispersed into small oil bubbles, i.e., oil-in-water bubble flow occurs. In this flow pattern, the oil phase exists in the form of discrete bubbles in water continuous phase.

When the total velocity of the oil-water mixture flow is high, very fine dispersed oil-in-water bubble flow gradually occurs. Due to the influence of high turbulent kinetic energy, the oil bubbles are broken into smaller oil droplets in the transition from oil-in-water bubble flow to very fine dispersed oil-in-water bubble flow. The motions of large numbers of smaller oil droplets become rather stochastic (Fig. 2.4).

References

1. G.F. Hewitt, *Measurement of Two-phase Flow Parameters* (Academic Press, London, 1978)
2. J.G. Flores, X.T. Chen, C. Sarica, J.P. Brill, Characterization of oil-water flow patterns in vertical and deviated wells. SPE Prod. Facil. **14**, 102–109 (1999)

Chapter 3
The Experimental Flow Loop Facility and Data Acquisition

3.1 Vertical Gas-Water Two-Phase Flow Experiment

The gas-water two-phase flow experiment in a 125 mm diameter vertical upward transparent Plexiglas pipe was carried out in the multiphase flow loop of Tianjin University. The experimental mediums are air and tap water. The schematic illustration of the experimental flow loop facility is shown in Fig. 3.1. The measurement system consists of several parts: a vertical multi-electrode array (VMEA) conductance sensor that was specifically designed for the two-phase flow experiment [1], high speed VCR (Video Camera Recorder), signal generating circuit, signal modulating module, data acquisition device (PXI 4472 card, National Instruments), and signal analysis software. The VMEA, as shown in Fig. 3.2, consists of eight alloy titanium ring electrodes axially separated and flush mounted on the inside wall of the flowing pipe. E1 and E2 are exciting electrodes. C1–C2 and C3–C4 are two pairs of upstream and downstream correlation electrodes denoted as sensor A and sensor B, respectively. Based on the cross-correlation technique, the axial velocity of the two-phase flow can be extracted from fluctuating signals through sensors A and B. H1–H2 is the volume-fraction electrodes denoted as sensor C. The measurement system uses a 20 kHz constant voltage (1.4 V) sinusoidal wave to excite the flow. The signal modulating module consists of three submodules: differential amplifier, sensitive demodulation, and low-pass filter. The data processing part is realized through graphical programming language LABVIEW 7.1 embedded in the data acquisition card, which can display, store, and analyze data wave forms in real time.

A typical experimental is implemented by generating water flow at a fixed rate in the pipe and then gradually increasing the gas-flow rate. When the gas and water flow rates reaches a pre-defined ratio, a conductance signal is collected from VMEA and the flow pattern is visualized by the high-speed VCR (the resolution and the frame rate was set at 640×480 and 200 frames s^{-1}, respectively). Figure 3.3 shows five flow patterns from the gas–water flows. The water- and gas-flow rates range from 0.02 to 0.27 m/s and from 0.0045 to 2.94 m/s, respectively. The sampling frequency is 400 Hz, and the data recording time for one

Z.-K. Gao et al., *Nonlinear Analysis of Gas-Water/Oil-Water Two-Phase Flow in Complex Networks*, SpringerBriefs on Multiphase Flow, DOI: 10.1007/978-3-642-38373-1_3, © The Author(s) 2014

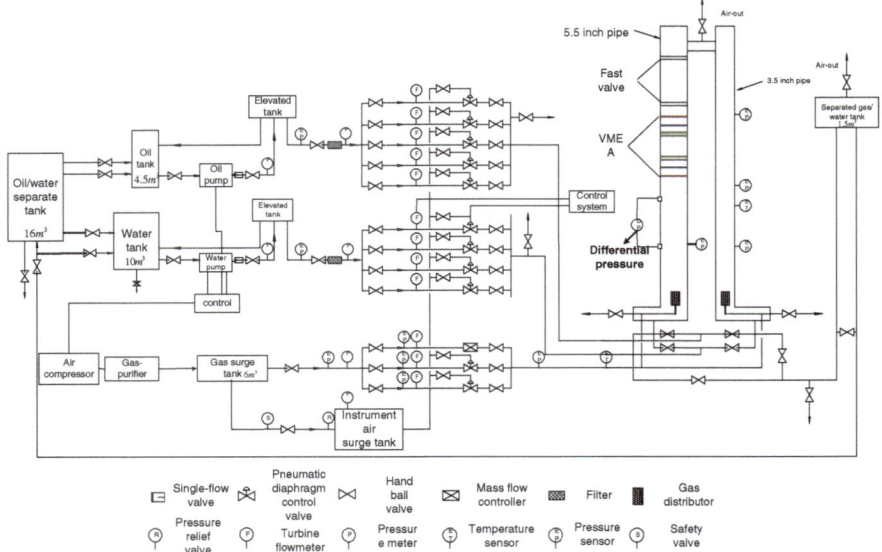

Fig. 3.1 Configuration of experimental flow loop facility

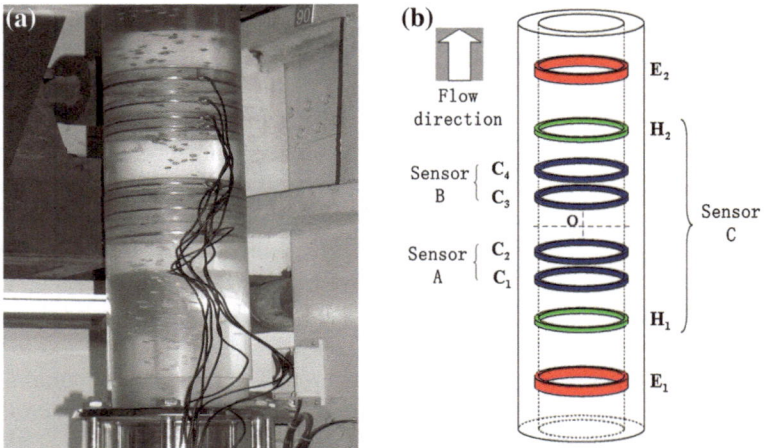

Fig. 3.2 The VMEA (Vertical Multi-Electrode Array) conductance sensor. **a** The VMEA measurement section; **b** The geometry of VMEA

measurement point is 60 s. The conductance fluctuating signals of VMEA sensor (sensor C) corresponding to the five gas-water flow patterns are shown in Fig. 3.4, where U_{sg} and U_{sw} represent gas superficial velocity and water superficial velocity, respectively.

3.2   Horizontal Gas-Water Two-Phase Flow Experiment

Horizontal gas-liquid two-phase flow experiment in horizontal pipe of inner diameter 125 mm was carried out in multiphase flow loop facility of Tianjin University. The experimental mediums are gas and water. The experimental plan was like that, first we put a fixed water flow rate into the pipe, then we gradually increased the gas flow rate; every time when the gas and water flow rates reached a pre-defined ratio, we acquired one conductance fluctuating signal from VMEA, and visualized the flow pattern by high speed VCR. Three observed horizontal gas-liquid flow patterns are stratified flow, stratified wavy flow and slug flow. The snapshots for three typical horizontal gas-liquid flow patterns are shown in Fig. 2.2. The sampling frequency was 400 Hz, and the sampling data recording time for one measuring point was 50 s. The water- and gas-flow rates are between 1 and 14 m^3/h and from 0.2 to 120 m^3/h, respectively. The conductance fluctuating signals of three typical flow patterns, measured from sensor C, are shown in Fig. 3.5, in which U_{sg} and U_{sw} represent gas flow rate and water flow rate, respectively.

3.3   Inclined Oil-Water Two-Phase Flow Experiment

The inclined oil-water two-phase flow experiment in a 125 mm diameter pipe was also carried out through the multiphase flow loop in Tianjin University. The transparent length of the Plexiglas pipe is 6 m and the pipe can be tilted at any angle from vertical (0°) to horizontal (90°). The flow rates of oil and water phase can be measured by calibrated turbine and electromagnetic flow meter. Figure 3.6a shows the experimental design of inclined oil-water flows. Compared to the gas-water flow, the oil-water flow belongs to liquid-liquid mixing flow so that it is rather difficult to obtain its flow pattern transition by the high-speed VCR. We thus exploit mini-conductance probes [2–4] to investigate different water-dominated countercurrent flows observed in the experiment.

The measurement system, as shown in Fig. 3.6, consists of the following parts: a vertical multi-electrode array (VMEA) conductance sensor, mini-conductance probes, signal generating circuit, signal modulating module, data acquisition device (PXI 4472 and 6115 card, National Instruments), and signal analysis software. Figure 3.7a shows the geometry structure of the mini-conductance probes. We can see that, the mini-conductance probes are mainly composed of needle electrode and conductive casing, the space between which is filled with insulation. Figure 3.7b shows the installation scheme of the mini-conductance probes, in which five mini-conductance probes are arrayed in equidistance at test section, and the distance between adjacent probes is 150 mm. Probe A, B, C, D, E are respectively at 1/8D, 1/8D, 1/4D, 1/4D, 1/2D from pipe wall, where D = 125 mm is the pipe diameter. When the flow loop is inclined from vertical, probes A and C are in the upper part of the pipe, probe E is in middle region and

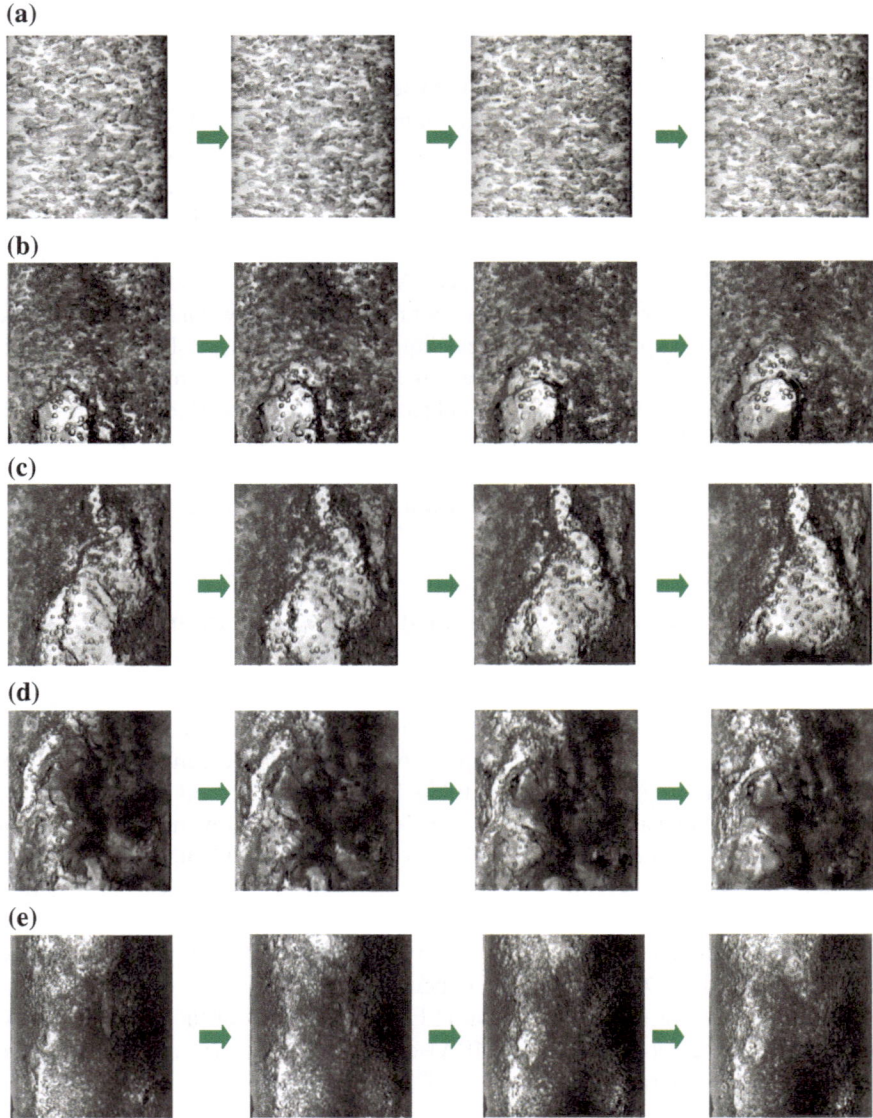

Fig. 3.3 Dynamic images of five gas-water flow patterns recorded by High-speed VCR. **a** The dynamic images of Bubble flow ($U_{sw} = 0.091$ m/s, $U_{sg} = 0.005$ m/s), **b** The dynamic images of Bubble-Slug transitional flow ($U_{sw} = 0.091$ m/s, $U_{sg} = 0.018$ m/s), **c** The dynamic images of Slug flow ($U_{sw} = 0.091$ m/s, $U_{sg} = 0.315$ m/s), **d** The dynamic images of Slug-Churn transitional flow ($U_{sw} = 0.091$ m/s, $U_{sg} = 0.566$ m/s), **e** The dynamic images of Churn flow ($U_{sw} = 0.091$ m/s, $U_{sg} = 1.517$ m/s)

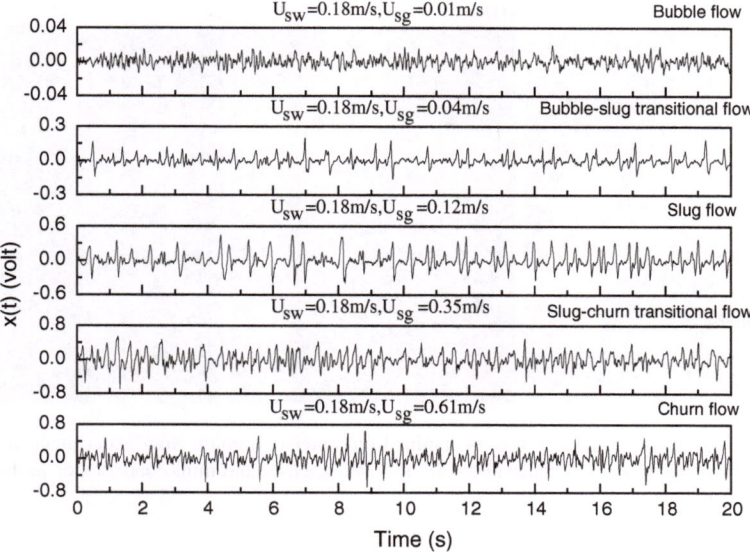

Fig. 3.4 The VMEA conductance fluctuating signals in five gas-water flow patterns

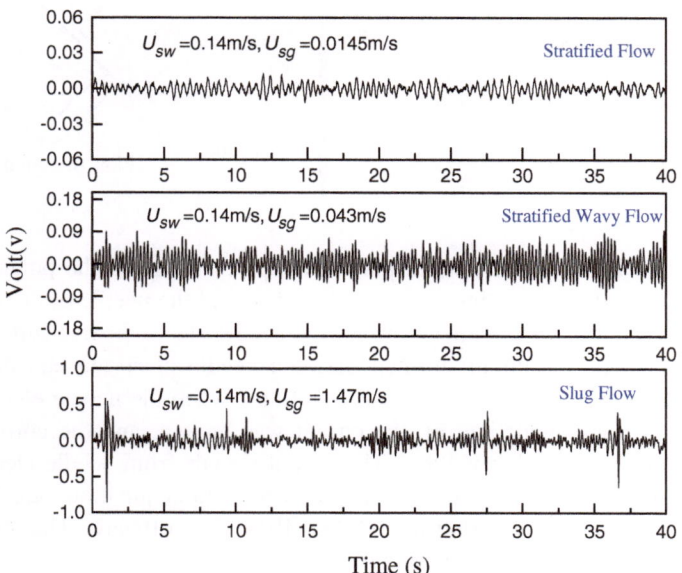

Fig. 3.5 Conductance fluctuating signals of three typical horizontal gas-liquid flow patterns

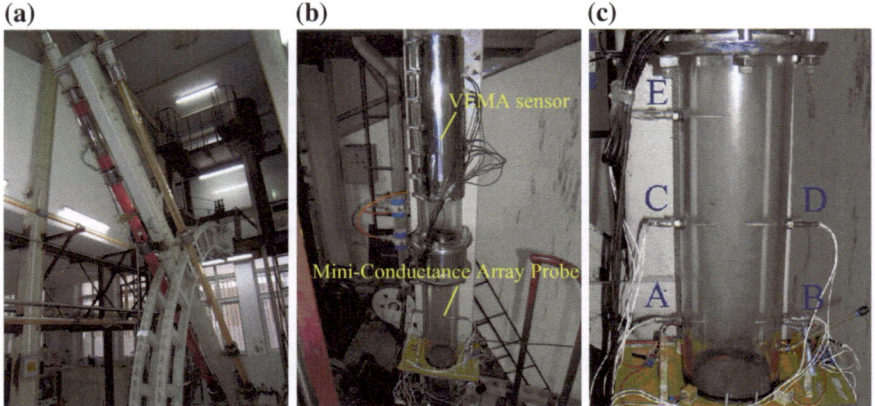

Fig. 3.6 The measurement system of inclined oil-water two-phase flow experiment. **a** Experimental setup of inclined oil-water flows; **b** the VMEA sensor and mini-conductance probes; **c** the mini-conductance probes

Fig. 3.7 The mini-conductance probes.
a Geometry structure of mini-conductance probes;
b installation schematics of mini-conductance probes

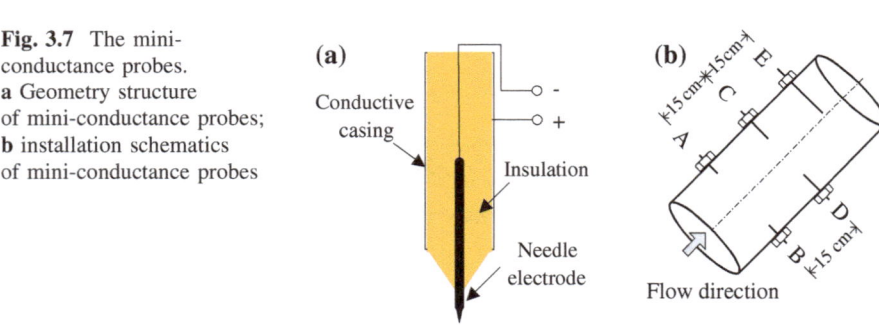

probes B and D are at the bottom part of the pipe. To avoid the polarization of needles, the conductive casing is excited by +5 V and the needle electrodes act as measurement port. In experiments, when the needle electrode was surrounded by conductive water, the circuit formed between needle electrode and casing, the corresponding output signals were at low voltage; when the needle electrode was immerged in non-conductive oil, the circuit was broken, and the corresponding output signals exhibit high voltage. The output signals from needle electrode are modulated by a follower circuit and acquired by data acquisition card PXI 4472 and PXI 6115 under the control of LABVIEW 7.1 software. The modulating circuit of mini-conductance probe is shown in Fig. 3.8.

The experimental mediums are tap water and No. 15 industry white oil, where the oil density is 856 kg/m^3, the viscosity is 11.984 mPa s (45 °C) and the surface tension is 0.035 N/m. The water phase is dyed with red colored for the convenience of visualization and the inclination angles is 45° from vertical. Experiments were implemented by initially injecting both water and oil flows into the pipe. The

Fig. 3.8 The modulating circuit of mini-conductance probe

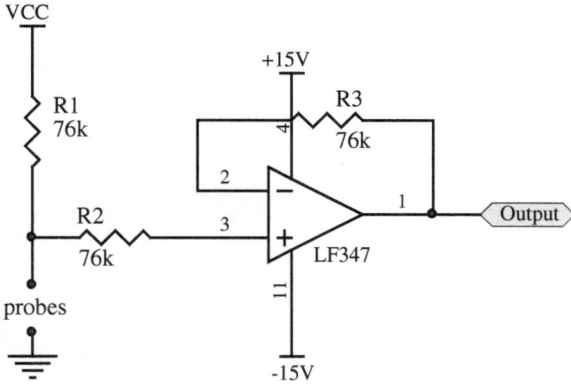

water flow rate was fixed and the oil flow rate was gradually increased. Each time when the ratio between the oil and water flow rates reached a certain preset value, conductance fluctuating signals were collected from the VMEA sensor and the mini-conductance probes. In our experiment, the water phase flow rate was set between 0.0057 and 0.3306 m/s and the oil phase rate ranges from 0.0052 to 0.3306 m/s. The sampling frequency was 400 Hz and the sampling time between two measuring points was 30 s. We have observed three different water-dominated inclined oil–water flow patterns in the experiment, as shown in Fig. 2.3, i.e., CT, PS, and transitional flow patterns. The conductance fluctuating signals of VMEA sensor (sensor C) corresponding to three water-dominated oil-water flow patterns are shown in Fig. 3.9, where U_{so} and U_{sw} represent oil superficial velocity and water superficial velocity, respectively. It should be pointed out that, because of the significant difference in electrical sensibility between gas/oil phase and water phase, the random flow of gas/oil phase will induce voltage fluctuation on the measuring electrode under a certain sinusoidal input, which implies that the conductance fluctuating signals measured from the VMEA conductance sensor are related to the flow transition. Based on the conductance fluctuating signals, we construct complex networks to study the gas-water/oil-water two-phase flow through analyzing the resulting networks.

The typical mini-conductance probe signals in different flow patterns in terms of increasing oil superficial velocity (U_{so}) at fixed water superficial velocity (U_{sw}) of 0.0374 m/s are shown in Fig. 3.10. Figure 3.10a shows the typical signals of mini-conductance probes in D O/W PS flow pattern. In this flow pattern, oil phase in the form of intermittent swarms move fast upward in the upper side of the pipe, while countercurrent water flow exists at the bottom of the pipe. The intermittent positive fluctuations from low to high voltage observed by probe A indicate the existence of intermittent oil swarms. Less intermittent positive fluctuations from probe C correspond to weaker intermittent oil swarms from the upper side to the lower side of the pipe. The signals of probes B, D and E, are all at low voltage and almost unchanged, indicating that countercurrent water exist in these locations

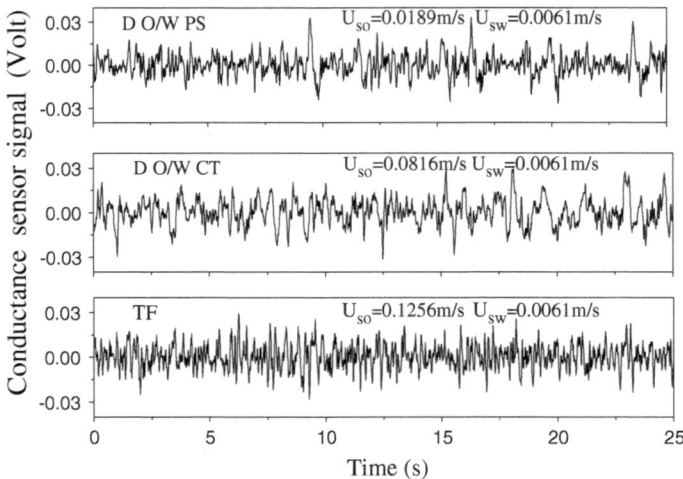

Fig. 3.9 The VMEA conductance fluctuating signals in three water-dominated oil-water flow patterns

instead of oil phase. Figure 3.10b shows the typical probe signals in D O/W CT flow pattern. In this flow pattern, the sequence of oil droplets observed in the countercurrent flow pattern is interrupted by water breaks, associated with the accumulation and packing of oil droplets at the top of the pipe. Probe A shows the alternatively occurrence of local oil-in-water and water-in-oil flow patterns in this flow conditions. This phenomenon also occurs at probe C, where more oil-in-water and less water-in-oil flow patterns arise. Probe E detects the existence of some oil droplets at the central of the pipe. Both probes B and D are at low voltage, demonstrating the occurrence of continuous water phase at the bottom of the pipe. Figure 3.10c shows typical probe signals in TF flow pattern. In this unsteady transitional flow pattern, oil swarms will coalesce into larger swarms, and due to the influences of agglomeration, coalescence and gravity components, thin and elongated oil film will form at the top of the pipe; Below the oil film, there exists a switch between oil-dominated and water-dominated flow patterns; while the local flow pattern at the bottom of the pipe is still oil-in-water flow. The signals of probes A and C display typical negative pulses from high voltage to low voltage, indicating the existence of water-in-oil flow pattern at the upper side of the pipe. In TF flow pattern, since probes A and C are typically surrounded by oil phase, their signals always keep at high voltage; once some dispersed water droplets touch the probe, the signals will change from high to low; once the water droplets leave the probes, the signals will turn back to high voltage again. Probe E, at the middle region of the pipe, shows that oil-in-water flow and water-in-oil flow patterns alternatively occurs with the predominant of local oil-in-water flow pattern. Probe D demonstrates continuous water flow associated with occasionally rise of oil droplets; while probe B still displays low voltage, indicating continuous water flow

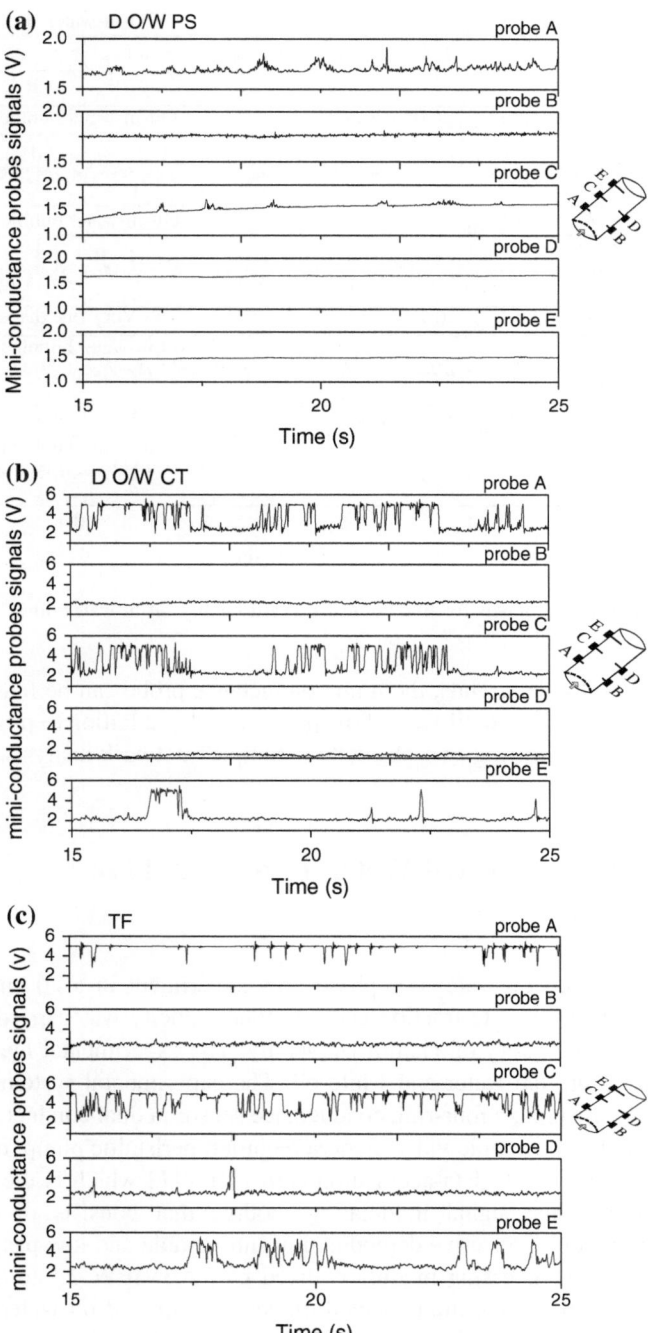

Fig. 3.10 The mini-conductance probe signals in three water-dominated oil-water flow patterns. **a** D O/W PS ($U_{so} = 0.0375$ m/s, $U_{sw} = 0.0374$ m/s); **b** D O/W CT ($U_{so} = 0.1682$ m/s, $U_{sw} = 0.0374$ m/s); **c** TF ($U_{so} = 0.1905$ m/s, $U_{sw} = 0.0374$ m/s)

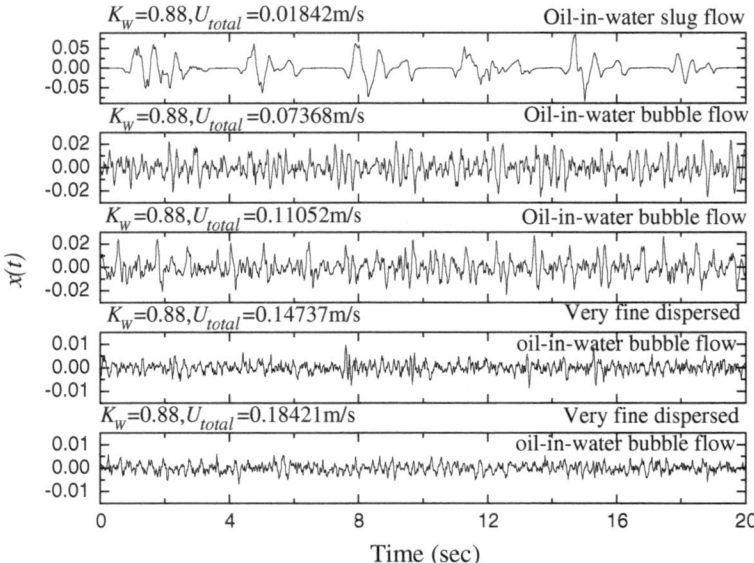

Fig. 3.11 Conductance signals for five vertical upward bubbly oil-in-water flows

without any oil droplets. Thus, the mini-conductance probe can be faithfully used to define different inclined oil-water flow patterns and in addition to provide useful reference for the following two-phase flow complex network analysis.

3.4 Vertical Upward Oil-Water Two-Phase Flow Experiment

The vertical upward oil-water two-phase flow experiment in a 20 mm diameter pipe at high water flowrate fraction and low total velocity was carried out in the two-phase flow laboratory of Tianjin University. The experimental mediums, i.e., the two phases, are tap water and white oil. The experimental system includes a vertical plexiglass pipe; a four-ring conductance sensor; a dual conductance probe; a digital high-speed dynamic video camera recorder; peristaltic pump; an exciting-signal generating circuit; data-acquisition card PXI 4472 which is the product of National Instruments; signal modulating module that consists of differential amplifier sub-module, sensitive demodulation sub-module and low-pass filter sub-module; signal preprocessing module realized by LABVIEW 7.1 to display and store data in real-time. We implemented the vertical upward oil-water two-phase flow experiments by initially injecting both water and oil flows into a 20 mm diameter pipe. We fixed the water flowrate fraction and then gradually increased the total velocity of the oil-water mixture flow. In the experiment, the water flowrate fraction was set at 80, 84, 88, 90 and 92 %, respectively, and the total velocity of the oil-water mixture flow was set at 0.01842, 0.03684, 0.07368, 0.11052, 0.14737, 0.18421 and 0.22105 m/s, respectively. We used the exciting-signal generating circuit to generate and input a 20 kHz sinusoidal voltage signal of amplitude 1.4 V to excite the oil-water concurrent flow. For one fixed water flowrate fraction, each time when the total velocity of the oil-water mixture flow reached a certain preset value, we collected conductance signals from the four-ring conductance sensor and dual conductance probe, respectively, and the flow pattern could be visualized and defined by the digital high-speed dynamic video camera recorder. The sampling frequency was 4000 Hz and the measuring time for one measurement was 30 s. Note that the conductance signals from the dual conductance probe are mainly used to detect the size of the oil bubbles, and the conductance signals from the four-ring conductance sensor are very sensitive to the transition of flow patterns and allow conducting recurrence network analysis to investigate the dynamic flow behavior. In our experiment, with the increase of total velocity of the oil-water mixture flow, oil bubbles in the mixture flow became smaller and smaller, correspondingly, three types of vertical upward oil-in-water flow patterns in a small diameter pipe gradually appeared, i.e., oil-in-water slug flow, oil-in-water bubble flow and very fine dispersed oil-in-water bubble flow, as shown in Fig. 2.4. Figure 3.11 shows the conductance signals from the four-ring conductance sensor for five oil-water flow patterns displayed in Fig. 2.4, where K_w and U_{total} represent water flowrate fraction and total velocity of the oil-water mixture flow, respectively.

References

1. N.D. Jin, Z. Xin, J. Wang, Z.Y. Wang, X.H. Jia, C.P. Chen, Design and geometry optimization of a conductive probe with a vertical multiple electrode array for measuring volume fraction and axial velocity of two-phase flow. Meas. Sci. Technol. **19**, 045403 (2008)
2. Y.V. Fairuzov, P. Aernas-Medina, J. Verdejo-Fierro, R. Gonzalez-IsIas, Flow pattern transitions in horizontal pipelines carrying oil–water mixtures: fullscale experiments. J. Energy Resour. Technol. **122**, 169–176 (2000)
3. P. Angeli, G. Hewitt, F. Flow structure in horizontal oil-water flow. Int. J. Multiphase Flow **26**, 1117–1140 (2000)
4. G.P. Lucas, R. Mishra, Measurement of bubble velocity components in a swirling gas-liquid pipe flow using a local four-sensor conductance probe. Meas. Sci. Technol. **16**, 749–758 (2005)

Chapter 4
Community Detection in Flow Pattern Complex Network

4.1 Community Detection in Gas-Water Flow Pattern Complex Network

4.1.1 Flow Pattern Complex Network

Flow Pattern Complex Network (FPCN) [1], extracted from the conductance fluctuating signals, is an abstract network, in which each flow condition is represented by a single node and the edge is determined by the strength of correlation between nodes. Flow condition refers to the flow behavior under different proportions of gas flow rate and water flow rate in the pipe. Since we configured 90 different proportions of gas flow rate and water flow rate to obtain 90 conductance fluctuating signals in the gas-water two-phase flow experiment, there are 90 different flow conditions (i.e., the number of nodes contained in FPCN is 90), in which each node corresponds to one of these 90 conductance fluctuating signals.

Note that the correlation between two nodes characterizes the correlation between two corresponding conductance fluctuating signals. We now demonstrate how the strength of correlation between conductance fluctuating signals can be used to establish edges. With respect to the nonlinear characteristics of the gas-water two-phase flow, we first apply the method of Time-Delay Embedding [2] to process the conductance fluctuating signals. That is, we use C–C method [3] to calculate the delay time τ from 90 conductance fluctuating signals, respectively, and choose the proper τ that can maximize the FPCN modularity [4]. Then we extract six time-domain features and four frequency-domain features from each processed conductance fluctuating signal to conduct the characteristic vector (see the following Part A for details). Therefore, there are 90 characteristic vectors and each vector contains ten elements. For each pair of characteristic vectors, $\mathbf{T_i}$ and $\mathbf{T_j}$, the correlation coefficient can be written as:

Z.-K. Gao et al., *Nonlinear Analysis of Gas-Water/Oil-Water Two-Phase Flow in Complex Networks*, SpringerBriefs on Multiphase Flow, DOI: 10.1007/978-3-642-38373-1_4, © The Author(s) 2014

$$C_{ij} = \frac{\sum_{k=1}^{M} [\mathbf{T_i}(k) - \langle \mathbf{T_i} \rangle] \cdot [\mathbf{T_j}(k) - \langle \mathbf{T_j} \rangle]}{\sqrt{\sum_{k=1}^{M} [\mathbf{T_i}(k) - \langle \mathbf{T_i} \rangle]^2} \cdot \sqrt{\sum_{k=1}^{M} [\mathbf{T_j}(k) - \langle \mathbf{T_j} \rangle]^2}}, \tag{4.1}$$

where M is the dimension of the characteristic vector and $\langle \mathbf{T_i} \rangle = \sum_{k=1}^{M} \mathbf{T_i}(k) \Big/ M$, $\langle \mathbf{T_j} \rangle = \sum_{k=1}^{M} \mathbf{T_j}(k) \Big/ M$. The elements C_{ij} are restricted to the range $-1 \leq C_{ij} \leq 1$, where $C_{ij} = 1$, 0 and -1 correspond to perfect correlations, no correlations and perfect anti-correlations, respectively. C is a symmetric matrix and C_{ij} describes the state of connection between node i and j. Finally, choosing a critical threshold r_c (see the following Part B for details), the correlation matrix C can be turned into adjacency matrix A, the rules of which read:

$$A_{ij} = \begin{cases} 1, & (|C_{ij}| \geq r_c) \\ 0, & (|C_{ij}| < r_c) \end{cases}, \tag{4.2}$$

Which means there will be an edge connecting nodes i and j, if $|C_{ij}| \geq r_c$. While there will not be an edge connecting node i and node j if $|C_{ij}| < r_c$. All the nodes and edges constitute the FPCN, and the topological structure of this network can be described by the adjacency matrix A.

(a) Extracted feature from conductance fluctuating signals

We extract ten different kinds of feature quantities in both time and frequency domains. In the time domain, we choose the maximum value, minimum value, average value, standard deviation, asymmetry coefficient and kurtosis function as the features of signals; in the frequency domain, we select the four coefficients of the linear prediction model with four orders.

Maximum value

$$x_{max} = \max(x_1, x_2, \ldots, x_n) \tag{4.3}$$

Minimum value

$$x_{min} = \min(x_1, x_2, \ldots, x_n) \tag{4.4}$$

Mean value

$$\bar{x} = \frac{1}{n-1} \sum_{i=1}^{n} x_i \tag{4.5}$$

Standard deviation

$$x_{sta} = \left(\frac{\sum_{i=1}^{n} (x_i - \bar{x})^2}{n-1} \right)^{1/2} \tag{4.6}$$

Dissymmetry coefficient

$$x_{dis} = \frac{\sum\limits_{i=1}^{n} (x_i - \bar{x})^3}{(n-1) \cdot S^2} \tag{4.7}$$

Kurtosis coefficient

$$x_{kur} = \left(\frac{\sum\limits_{i=1}^{n} (x_i - \bar{x})^4}{(n-1) \cdot S^4} \right) - 3 \tag{4.8}$$

The method of feature extraction in the frequency domain was mainly referred to that used by Darwich et al. [5] which was based on the concept of linear prediction in speech signal processing [6]. The basic principle of this method is that the value of a signal point at present could be estimated by the linear combination of several previous points; the coefficient of this linear combination could be acquired when the deviation between the true value and the estimated value is minimum. The coefficients and previous signals comprise a linear prediction model; the coefficients are called the orders of this model, which are also what we set as the frequency characteristic quantities. We assumed that the input I_t could be expressed as follows:

$$I_t = -\sum_{k=l}^{p} c_k Z_{t-k} + G \sum_{l=0}^{q} d_l U_{t-l} \quad (d_0 = 1) \tag{4.9}$$

where U_{t-l} are the unknown input signals, c_k ($1 \le k \le p$), d_l ($1 \le l \le q$) and G are all system parameters, of which G is the system gain, and c_k ($1 \le k \le p$), the coefficients of the linear combination, are the frequency characteristic quantities. We choose the linear prediction model with four orders, so that c_1, c_2, c_3, c_4 are needed to be figured out. The detailed computation process can be seen in [6].

(b) Selection of the threshold

Before exploring how to select a critical threshold r_c, here we first introduce a quality function or "modularity" Q, proposed by Newman et al. [4]. Let e_{ij} be the fraction of edges in the network that connect nodes in community i to those in community j, and let $a_i = \sum_j e_{ij}$. Then

$$Q = \sum_i (e_{ii} - a_i^2) \tag{4.10}$$

is the fraction of edges within communities, minus the expected value of the same quantity in the absence of community structures. If a particular partition yields no more within-community edges than that would be expected by the random counterpart, we will get $Q = 0$. Values other than 0 indicate deviations from

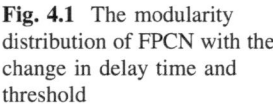

Fig. 4.1 The modularity
distribution of FPCN with the
change in delay time and
threshold

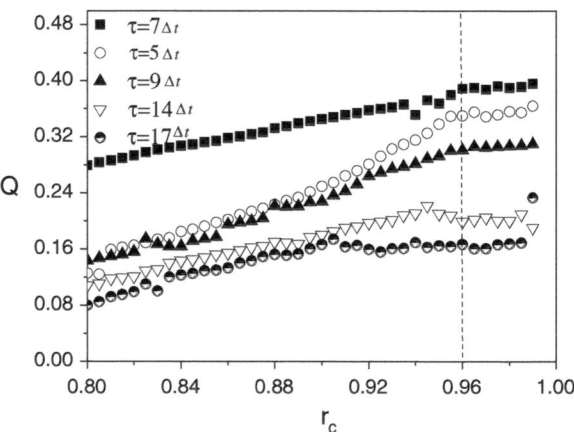

randomness, and in particular, values greater than about 0.3 suggest significant
community structure [4].

The threshold r_c determines the connections of the resulting network. If r_c is
extremely small, the pairs of nodes with weak correlations will be connected as
well. In this scenario, the physical content of correlations in time series will be
hidden. Increasing the value of r_c will reduce the number of connections among
nodes and more noise will be filtered out. But so far, there has not been a general
method for determining the critical threshold. In this study, we can estimate a
range of r_c, in which the structure of the resulting network is relatively stable. In
particular, a critical threshold r_c can be found just by simulating a specific process
of the complex network, i.e., decreasing the number of connections by increasing
the value of r_c while keeping the modularity of resulting network almost
unchanged. If there exists a neighborhood of a threshold, in which the modularity
of the resulting network has been almost the same, we can claim that the threshold
r_c is figured out.

The modularity distribution of FPCN with the change in delay time and
threshold is shown in Fig. 4.1. We see that, the modularity Q is stable when r_c
ranges from 0.965 to 0.985 and the modularity Q reaches the maximum when
$\tau = 7\Delta t$ (Δt is the sample interval of the conductance fluctuating signals).
Therefore, according to the above analysis, we choose $\tau = 7\Delta t$ and $r_c = 0.975$ to
establish the FPCN.

4.1.2 Flow Pattern Identification in Flow Pattern Complex Network

(a) Community-detection algorithm based on K-means clustering

Community-detection algorithm based on K-means clustering, proposed in this
paper is realized by using K-means approach to do clustering analysis on the data

which are obtained through conversion by Capocci's approach [7] to reveal the network community structure.

Capocci approach: Capocci recasted the eigenproblem into an optimization problem. Let the standard matrix be defined as:

$$H = K^{-1}A \qquad (4.11)$$

Consider the following constrained optimization problem:
Let $z(x)$ be defined as

$$z(x) = \frac{1}{2} \sum_{i,j=1}^{n} (x_i - x_j)^2 A_{ij} \qquad (4.12)$$

where x_i are values assigned to the nodes, with some constraint on the vector X, expressed by

$$\sum_{i,j=1}^{n} x_i x_j M_{ij} = 1 \qquad (4.13)$$

where M_{ij} are elements of a given symmetric matrix M. The stationary points of z over all X subject to constraint (4.12) are the solutions of

$$(D - A)X = \mu MX \qquad (4.14)$$

where D is the diagonal matrix :

$$\begin{cases} D = (d_{ij}) \\ d_{ij} = \delta_{ij} \sum_{k=1}^{n} A_{ik} \end{cases} \qquad (4.15)$$

A is the adjacency matrix, n is the number of nodes and μ is a Lagrange multiplier.

Different choices of the constraint M leads to different eigenvalues problems: for example choosing $M = D$ leads the eigenvalues problem $D^{-1}AX = (1 - 2\mu)X$, while $M = 1$ leads to $(D - A)X = \mu X$. Thus, $M = D$ and $M = 1$, corresponds to the eigenproblems for the (generalized) Normal and Laplacian matrix, respectively.

Capocci et al. [7] have proved that, for a network with significant community structure, the first non-trivial eigenvector components of the standard matrix H could be used to study its community structure. Therefore, we rely on the eigenvectors of standard matrix to conduct the clustering analysis and then derive the corresponding nodes to reveal the community structure.

(b) Flow pattern identification based on community structure

After constructing the FPCN containing 90 nodes, we show in Fig. 4.2 the distribution of the corresponding elements of the three first non-trivial eigenvectors. Three different communities are clearly identified when the components of

Fig. 4.2 Components of the
first non-trivial eigenvector a_1
are plotted versus those of a_2
and a_3

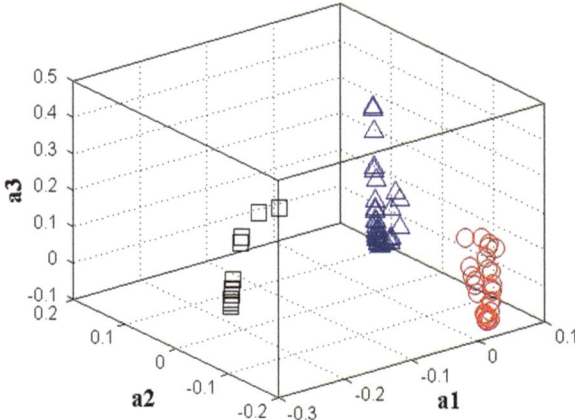

the first non-trivial eigenvector a_1 are plotted versus those of a_2 and a_3. The community structure of the FPCN, detected by community-detection algorithm based on K-means clustering, is shown in Fig. 4.3. The community structure is drawn by the software "Ucinet" and "Netdraw".

From the detected community structure, as shown in Fig. 4.3, three communities of 21, 30 and 39 nodes are found, denoted as community a, community b and community c, respectively. In addition, combining with the experimental observations using high speed VCR, we find that community a mainly corresponds to bubble flow, such as node 2 ($Q_g = 0.2$ m^3/h, $Q_w = 2.0$ m^3/h) and node 16 ($Q_g = 0.94$ m^3/h, $Q_w = 12.0$ m^3/h) both corresponding to bubble flow; community b mainly corresponds to slug flow, such as node 31 ($Q_g = 2.1$ m^3/h, $Q_w = 2.0$ m^3/h) and node 44 ($Q_g = 4.1$ m^3/h, $Q_w = 6.0$ m^3/h) both corresponding to slug flow; community c mainly corresponds to churn flow, such as node 70 ($Q_g = 69.0$ m^3/h, $Q_w = 4.0$ m^3/h) and node 90 ($Q_g = 139.0$ m^3/h, $Q_w = 2.0$ m^3/h) both corresponding to churn flow; the nodes of the FPCN that connect tightly between community a and community b correspond to bubble-slug transitional flow, such as node 19 ($Q_g = 1.0$ m^3/h, $Q_w = 2.0$ m^3/h) and node 26 ($Q_g = 1.7$ m^3/h, $Q_w = 4.0$ m^3/h) both corresponding to the bubble-slug transitional flow; the nodes of the FPCN that connect tightly between community b and community c correspond to slug-churn transitional flow, such as node 32 ($Q_g = 38.0$ m^3/h, $Q_w = 8.0$ m^3/h) and node 58 ($Q_g = 25.0$ m^3/h, $Q_w = 4.0$ m^3/h) both corresponding to the slug-churn transitional flow. In this regard, through detecting the community structure of the FPCN by the community-detection algorithm based on K-means clustering, we can successfully identify gas-water two-phase flow patterns by finding the three communities which correspond to the bubble flow, slug flow and churn flow respectively and the nodes that connect tightly between two communities corresponding to the transitional flow.

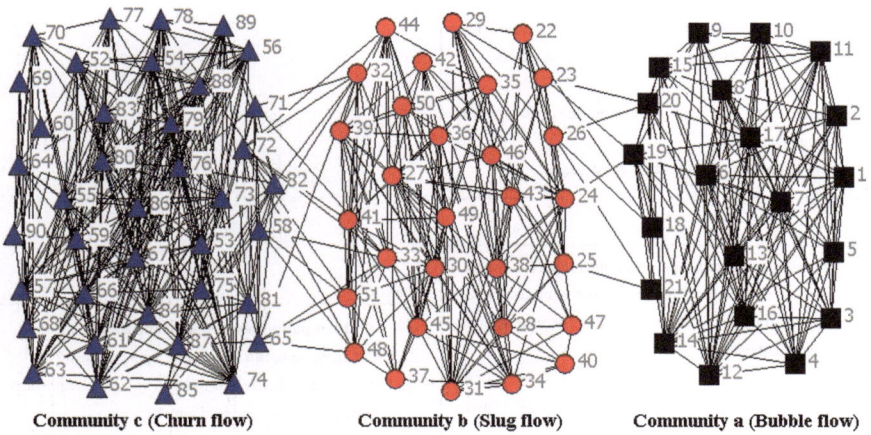

Community c (Churn flow)   Community b (Slug flow)   Community a (Bubble flow)

Fig. 4.3 Community structure of the gas-water FPCN

4.2   Community Detection in Oil-Water Flow Pattern Complex Network

We adopt the method mentioned above to construct the oil-water flow pattern complex network. The flow chart for detecting inclined oil-water flow pattern complex network is shown in Fig. 4.4. According to the principles mentioned in the above section, we choose $\tau = 9\Delta t$ (Δt is the sample interval of the conductance fluctuating signals) and $r_c = 0.98$ to establish the oil-water FPCN. After constructing the FPCN containing 48 nodes, we detect its community structure by using the community-detection algorithm based on K-means clustering. Figure 4.5 shows the distribution of the corresponding elements of the three first non-trivial eigenvectors. Apparently, three different communities can be clearly identified when the components of the first non-trivial eigenvector $\mathbf{a_1}$ are plotted versus those of $\mathbf{a_2}$ and $\mathbf{a_3}$. The community structure of the FPCN, detected by community-detection algorithm based on K-means clustering, is shown in Fig. 4.6. The community structure is drawn by the software "Ucinet" and "Netdraw".

From the detected community structure, as shown in Fig. 4.6, three communities of 30, 11 and 7 nodes are found, denoted by community a, community b and community c, respectively. Associated with the experimental observations, we find that community a mainly corresponds to the Dispersion Oil in Water-Pseudoslugs (D O/W PS) flow, such as node 2 ($U_{so} = 0.0189$ m/s, $U_{sw} = 0.0061$ m/s) and node 10 ($U_{so} = 0.0377$ m/s, $U_{sw} = 0.0189$ m/s) both corresponding to D O/W PS flow; community b mainly corresponds to the Dispersion Oil in Water-Counter-current (D O/W CT) flow, such as node 32 ($U_{so} = 0.0816$ m/s, $U_{sw} = 0.0061$ m/s) and node 36 ($U_{so} = 0.2423$ m/s, $U_{sw} = 0.0698$ m/s) both corresponding to D O/W CT flow; community c mainly corresponds to the Transitional Flow (TF), such as node 42 ($U_{so} = 0.1256$ m/s, $U_{sw} = 0.0061$ m/s) and node 48 ($U_{so} = 0.2875$ m/s,

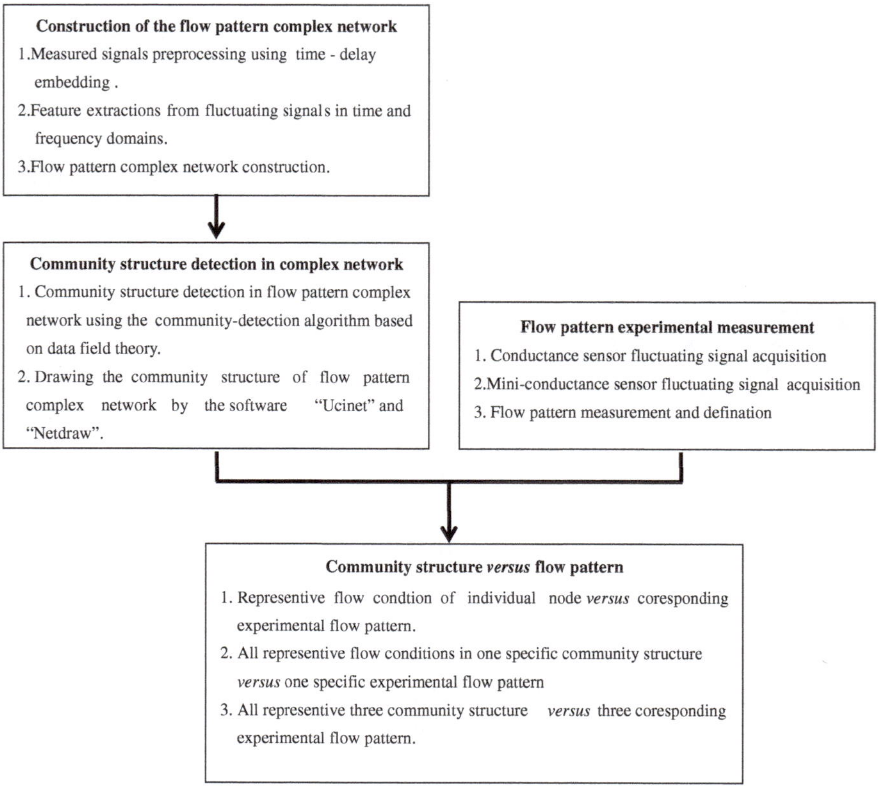

Construction of the flow pattern complex network

1.Measured signals preprocessing using time - delay embedding .

2.Feature extractions from fluctuating signals in time and frequency domains.

3.Flow pattern complex network construction.

Community structure detection in complex network

1. Community structure detection in flow pattern complex network using the community-detection algorithm based on data field theory.

2. Drawing the community structure of flow pattern complex network by the software "Ucinet" and "Netdraw".

Flow pattern experimental measurement

1. Conductance sensor fluctuating signal acquisition

2.Mini-conductance sensor fluctuating signal acquisition

3. Flow pattern measurement and defination

Community structure *versus* flow pattern

1. Representative flow condtion of individual node *versus* coresponding experimental flow pattern.

2. All representative flow conditions in one specific community structure *versus* one specific experimental flow pattern

3. All representative three community structure *versus* three coresponding experimental flow pattern.

Fig. 4.4 Flow chart for identifying inclined oil-water two-phase flow pattern based on community structure of flow pattern complex network

Fig. 4.5 Components of the first non-trivial eigenvector a_1 are plotted versus those of a_2 and a_3

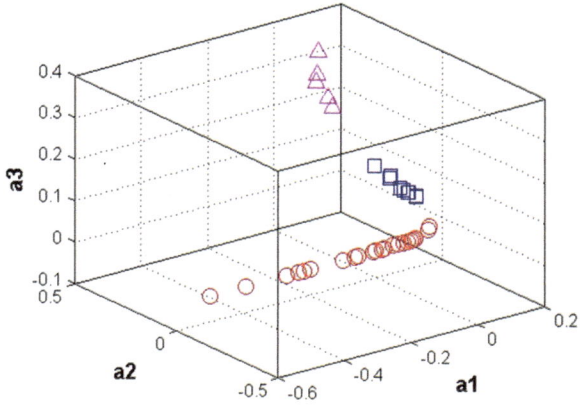

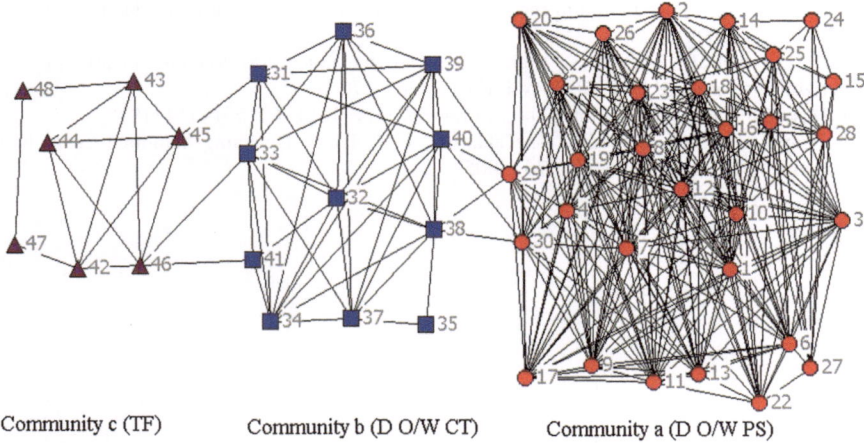

Community c (TF)   Community b (D O/W CT)   Community a (D O/W PS)

Fig. 4.6 Community structure of the oil-water FPCN

$U_{sw} = 0.0703$ m/s) both corresponding to transitional flow. Based on the community structure of flow pattern complex network, we also plot the oil-water flow pattern map for a 12.5 cm diameter pipe at an inclination angle of 45°. Hence, through detecting the community structure of the FPCN, we have successfully identified the inclined oil-water two-phase flow patterns by finding the three communities which correspond to the D O/W PS flow, D O/W CT flow and the transitional flow, respectively.

Note that, for the FPCN, we need to construct complex network and then using the community-detection algorithm detect the communities to identify different two-phase flow patterns. Although the processes seem a little complicated, they are necessary and effective. For example, for the inclined oil-water two-phase flow, it is rather difficult to distinguish D O/W CT flow from D O/W PS flow only by looking at the distributions of the experimental signals (see Fig. 3.13). With the help of FPCN, we can clearly identify PS, CT and TF flow pattern from the detected community structure, as shown in Fig. 4.6. Therefore, the flow pattern complex network could potentially be a powerful tool for distinguishing the two-phase flow patterns.

References

1. Z.K. Gao, N.D. Jin, Flow-pattern identification and nonlinear dynamics of gas-water two-phase flow in complex networks. Phys. Rev. E **79**(6), 066303 (2009)
2. N.H. Packard, J.P. Crutehfield, J.D. Farmer, Geometry from a time series. Phys. Rev. Lett. **45**(9), 712–716 (1980)
3. H.S. Kim, R. Eykholt, J.D. Salas, Nonlinear dynamics, delay times, and embedding windows. Physica D **127**, 48–60 (1999)

4. M.E.J. Newman, Fast algorithm for detecting community structure in networks. Phys. Rev. E **69**, 066133 (2004)
5. T.D. Darwich, H. Toral, J.S. Archer, Software technique for flow-rate measurement in horizontal two-phase flow. SPE Prod. Eng. **8**, 265–270 (1991)
6. J. Makhoul, Linear prediction: a tutorial review. Proc. IEEE **63**, 561–580 (1975)
7. A. Capocci, V.D.P. Servedio, G. Caldarelli, F. Colaiori, Detecting communities in large networks. Physica A **352**, 669–676 (2005)

Chapter 5
Nonlinear Dynamics in Fluid Dynamic Complex Network

5.1 Gas-Water Fluid Dynamic Complex Network

5.1.1 Construction of Fluid Dynamic Complex Network

To gain insight into the nonlinear dynamics of gas-water two-phase flow, we construct Fluid Dynamic Complex Network (FDCN) from one conductance fluctuating signal. Each segment of signal time series is represented by a single node and edges are determined by the strength of correlation between segments. Considering a conductance fluctuating signal (i.e., a time series), denoted as $\{z_1, z_2, z_3, \ldots, z_v\}$, we can obtain all the possible segments with length L, which read:

$$\{\mathbf{S}_1 = (z_1, z_2, \ldots, z_L)\}$$
$$\{\mathbf{S}_2 = (z_2, z_3, \ldots, z_{L+1})\}$$
$$\{\mathbf{S}_3 = (z_3, z_4, \ldots, z_{L+2})\} \tag{5.1}$$
$$\ldots$$
$$\{\mathbf{S}_w = (z_w, z_{w+1}, z_{w+2}, \ldots, z_{w+L-1}) | w = 1, \ldots, v - L + 1\}$$

For each pair of segments, $\mathbf{S_i}$ and $\mathbf{S_j}$, the correlation coefficient can be written as:

$$C_{ij} = \frac{\sum\limits_{k=1}^{L} \left[\mathbf{S_i}(k) \cdot \mathbf{S_j}(k) \right]}{\sqrt{\sum\limits_{k=1}^{L} \left[\mathbf{S_i}(k) \right]^2} \cdot \sqrt{\sum\limits_{k=1}^{L} \left[\mathbf{S_j}(k) \right]^2}} \tag{5.2}$$

C is a symmetric matrix and C_{ij} describes the state of connection between node i and j. Choosing a critical threshold r_c, the correlation matrix C can be converted into adjacent matrix A, the rules of which read: $A_{ij} = 1$ if $|C_{ij}| \geq r_c$ and $A_{ij} = 0$ if $|C_{ij}| < r_c$.

Z.-K. Gao et al., *Nonlinear Analysis of Gas-Water/Oil-Water*
Two-Phase Flow in Complex Networks, SpringerBriefs on Multiphase Flow,
DOI: 10.1007/978-3-642-38373-1_5, © The Author(s) 2014

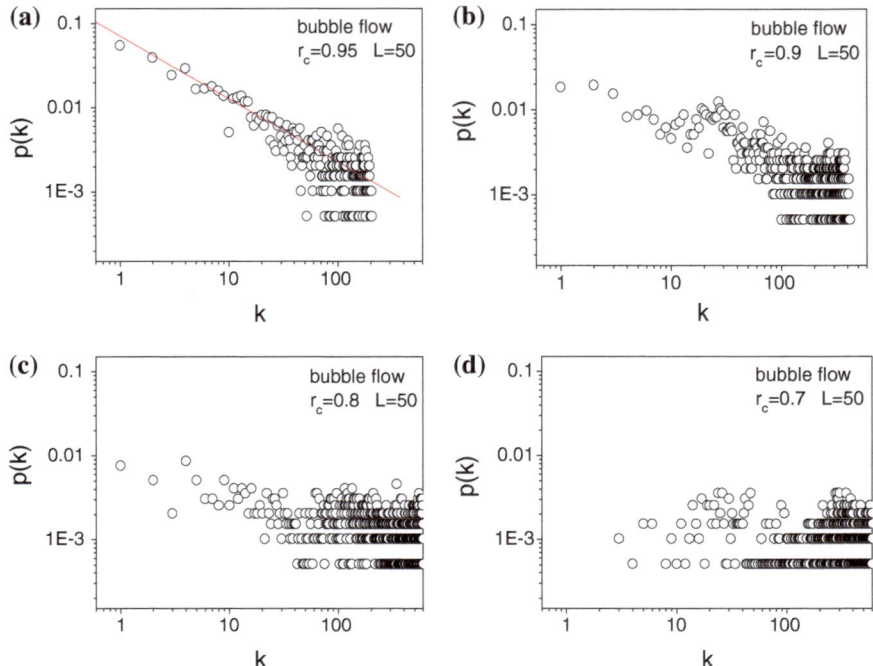

Fig. 5.1 Evolution of the gas-water FDCN constructed from the conductance fluctuating signal of bubble flow ($Q_g = 0.2$ m³/h, $Q_w = 2.0$ m³/h). When $r_c = 0.95$, the degree distribution obeys a power law. With the decrease of r_c, more and more edges are added, which may induce statistical fluctuations in the degree distribution. Finally, the power law is submerged in statistical noises. All the four figures are plotted in log-log scale

Rho et al. [1] have studied the characteristics of the degree distribution functions of the constructed networks. At a special critical threshold the degree distribution will tend to obey a power law. The network at this transition point is used to detect non-trivial characteristics embedded in the autocorrelation matrix. Through the analysis of the evolution of FDCN containing 2,000 nodes, we find that the resulting network can well keep the physically meaningful correlations when r_c is chosen as 0.95 (see Fig. 5.1 for details).

The other adjustable parameter is the length of a segment, denoted as L. Short length will induce overestimated correlations [2]. Increasing the length L can depress the finite-length-induced statistical fluctuations effectively. It should be sufficiently long to yield reliable results. As shown in Fig. 5.1, we find that when r_c and L are chosen as 0.95 and 50, respectively, the degree distribution of the resulting network can be well fitted with a power law as follows:

$$p(k) \sim k^{-\gamma} \tag{5.3}$$

where the degree (or connectivity) k of a node is the number of edges incident with it, and the degree distribution $p(k)$ is defined as the probability that a node chosen

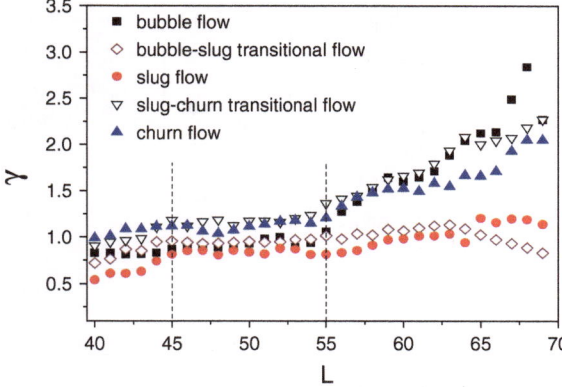

Fig. 5.2 The power-law exponent distribution of five gas-water FDCNs with the change in segment length. Bubble flow ($Q_g = 0.2$ m³/h, $Q_w = 2.0$ m³/h); Bubble-slug transitional flow ($Q_g = 1.0$ m³/h, $Q_w = 2.0$ m³/h); Slug flow ($Q_g = 4.1$ m³/h, $Q_w = 6.0$ m³/h); Slug-churn transitional flow ($Q_g = 25.0$ m³/h, $Q_w = 4.0$ m³/h); Churn flow ($Q_g = 139.0$ m³/h, $Q_w = 2.0$ m³/h)

uniformly at random has degree k; γ is the power-law exponent of degree distribution. In this study, we expect there is a stability region of L, in which the resulting network can reveal the physically meaningful information embedded in the time series. In particular, a proper L can be found through simulating a special dynamical process of the complex network, i.e., changing the number of nodes and connections by increasing the value of L, which does not change the power-law exponent of resulting network. Figure 5.2 displays the distribution of power-law exponent of different flow patterns. We see that the power-law exponent becomes relatively stable when L ranging from 45 to 55. We thus choose immediate value of $L = 50$ and $r_c = 0.95$ to derive the gas-water FDCN.

5.1.2 Network Degree Distribution and Its Physical Implications

By considering segments as nodes and defining network connectivity to be the correlation among segments, dynamics in time domain are naturally encoded into a network configuration. We select conductance fluctuating signals from five types of flow pattern for constructing five gas-water FDCNs, each of which contains 2,000 nodes. We find that the degree distributions of five networks can be well fitted by a power law, which indicates that the gas-water FDCNs are scale-free networks (see Fig. 5.3 for details).

To reveal the physical implications of the degree distribution of gas-water FDCNs, the method of recurrence plot [3] is employed. The recurrence plots of the

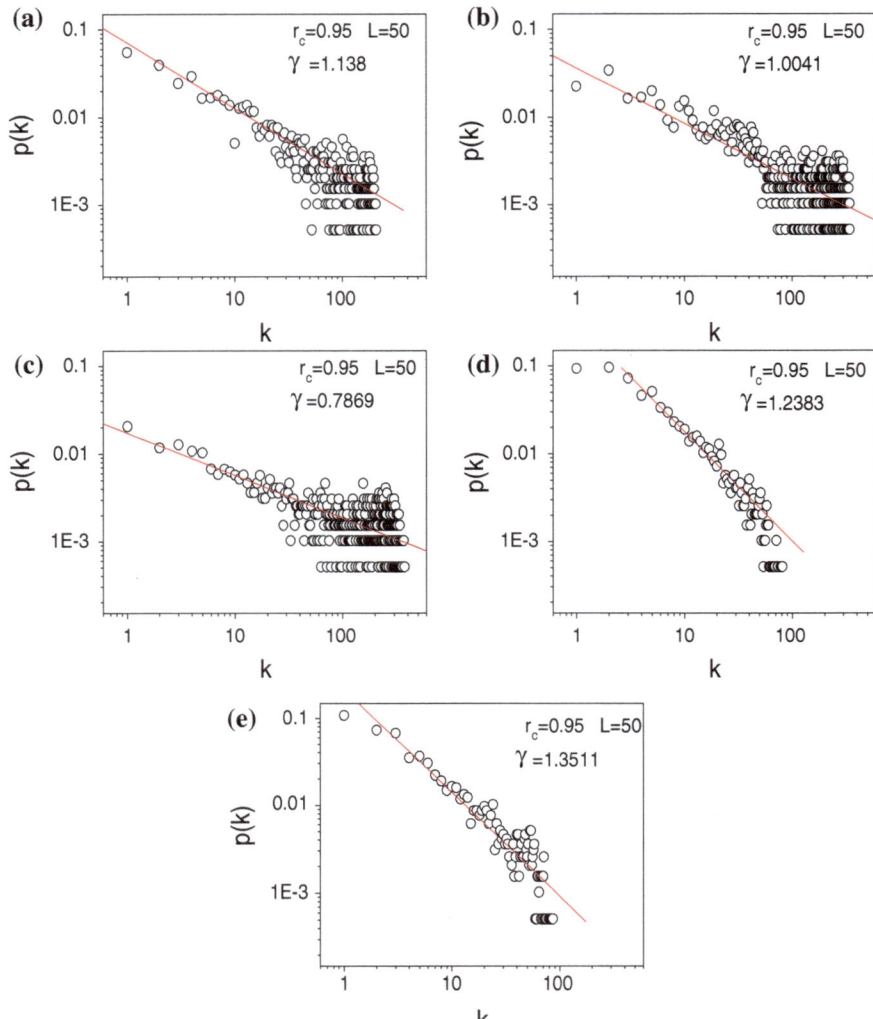

Fig. 5.3 Degree distribution of different types of gas-water FDCNs in log-log scale. **a** Bubble flow ($Q_g = 0.2$ m³/h, $Q_w = 2.0$ m³/h); **b** Bubble-slug transitional flow ($Q_g = 1.0$ m³/h, $Q_w = 2.0$ m³/h); **c** Slug flow ($Q_g = 4.1$ m³/h, $Q_w = 6.0$ m³/h); **d** Slug-churn transitional flow ($Q_g = 25.0$ m³/h, $Q_w = 4.0$ m³/h); **e** Churn flow ($Q_g = 139.0$ m³/h, $Q_w = 2.0$ m³/h)

five typical conductance fluctuating signals are shown in Fig. 5.4. The stochastic motion of large number of small bubbles can be reflected by the texture of its recurrence plot (i.e., large number of homogeneous discrete points) (see Fig. 5.4a). Along with the increase of gas superficial velocity, due to the periodic alternating movement between gas plug and liquid plug, the slug flow conductance fluctuating signal exhibits the periodic property to some extent, and this property can be

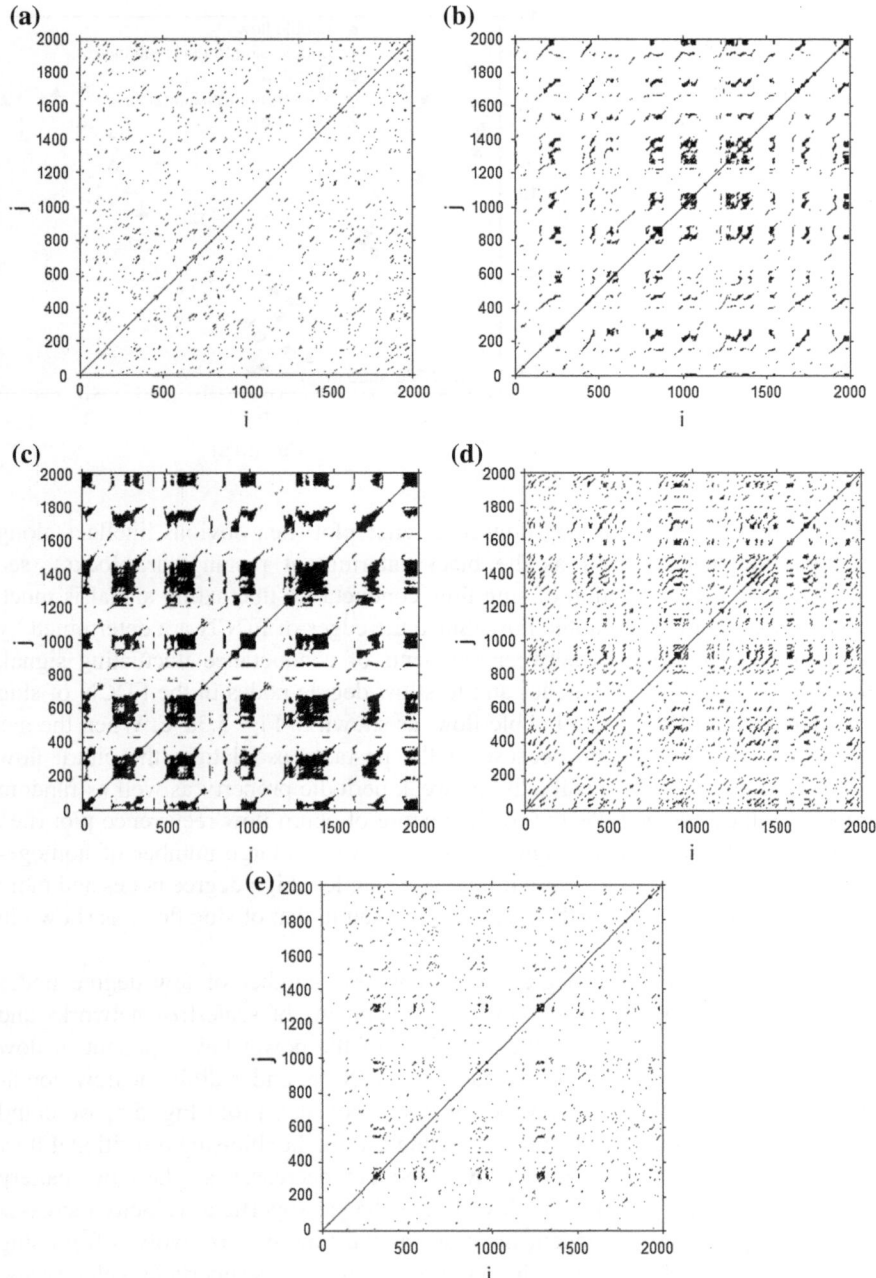

Fig. 5.4 Texture of recurrence plot of five gas-water flow patterns. **a** Bubble flow ($Q_g = 0.2$ m^3/h, $Q_w = 2.0$ m^3/h); **b** Bubble-slug transitional flow ($Q_g = 1.0$ m^3/h, $Q_w = 2.0$ m^3/h); **c** Slug flow ($Q_g = 4.1$ m^3/h, $Q_w = 6.0$ m^3/h);   **d**   Slug-churn   transitional   flow   ($Q_g = 25.0$ m^3/h, $Q_w = 4.0$ m^3/h); **e** Churn flow ($Q_g = 139.0$ m^3/h, $Q_w = 2.0$ m^3/h)

Fig. 5.5 The power-law exponent distribution in semi-log scale of 50 gas-water FDCNs with the increase of gas superficial velocity

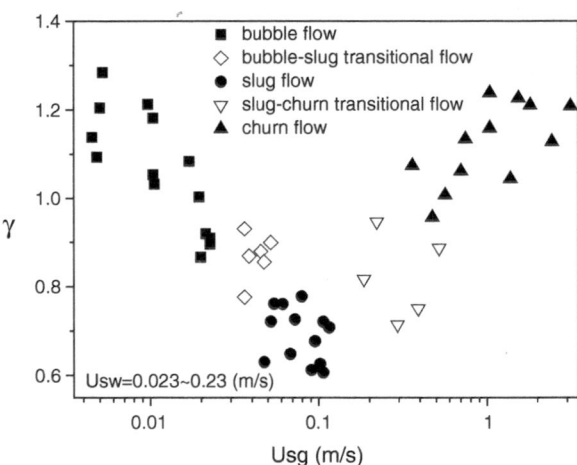

reflected by the texture of slug flow recurrence plot (i.e., obvious linellaes along principal diagonal as well as the black intermittent rectangular blocks) (see Fig. 5.4c). So the correlation of slug flow conductance fluctuating signal is much stronger than that of the bubble flow. Since the edges of FDCN are determined by the strength of correlation between segments of conductance fluctuating signal, there are more high degree nodes and less low degree nodes in the FDCN of slug flow than in the FDCN from bubble flow, as shown in Fig. 5.3a–c. When the gas superficial velocity is high, because of the unstable oscillation, the churn flow conductance fluctuating signal exhibits weak periodic property as well as random property, which can be reflected by the texture of churn flow recurrence plot (i.e., small black intermittent rectangular blocks as well as large number of homogeneous discrete points) (see Fig. 5.4e). So there are less high degree nodes and more low degree nodes in the FDCN of churn flow than in that of slug flow, as shown in Fig. 5.3d, e.

Larger number of high degree nodes and less number of low degree nodes correspond to lower values of power-law exponents of scale-free networks and vice versa. To further explore the variations of the power-law exponent in flow pattern transition, we construct 50 gas-water FDCNs under different flow conditions and calculate their relevant power-law exponents. From Fig. 5.5, we could see that, the power-law exponents of bubble flow and bubble-slug transitional flow are usually large, and the power-law exponent decreases as the flow pattern evolves from bubble flow to slug flow. But as the gas superficial velocity increases further, the power-law exponent increases as the flow pattern evolves from slug flow to churn flow. U_{sg} and U_{sw}, in Fig. 5.5, represents gas superficial velocity and water superficial velocity, respectively.

5.1.3 Network Information Entropy

The concept of Shannon's entropy [4] is the central role of information theory referred as measure of uncertainty. The entropy of a random variable is defined in terms of its probability distribution and can be shown to be a good measure of uncertainty. To calculate the information entropy I, Shannon also gave the equation as follows:

$$I = k_B \ln \Omega = -\sum_{j=1}^{n} k_B P(j) \ln P(j) \tag{5.4}$$

where Ω is the information, $P(j) = 1/\Omega$ and k_B is the Boltzmann constant.

According to the definition of Shannon's entropy, we define network information entropy for the FDCN and apply to investigate the nonlinear dynamics of gas-liquid flow through analyzing the network information entropy.

Definition 1: Let $P(i)$ be the importance of node i:

$$P(i) = k_i \bigg/ \sum_{j=1}^{n} k_j \tag{5.5}$$

where n is the number of nodes contained in the network and $k_i(k_j)$ is the degree of node $i(j)$.

Definition 2: Let E be the network information entropy:

$$E = -\sum_{i=1}^{n} k_B P(i) \ln P(i) \tag{5.6}$$

where n is the number of nodes contained in the network and k_B is the Boltzmann constant. In order to simplify the calculation, here we let $k_B = 1$, such that:

$$E = -\sum_{i=1}^{n} P(i) \ln P(i) \tag{5.7}$$

In order to make the network information entropy independent of the number of nodes contained in the network, we normalize E as follows:

$$E_N = \frac{E - E_{\min}}{E - E_{\max}} = \frac{-\sum_{i=1}^{n} P(i) \ln P(i) - \frac{\ln 4(n-1)}{2}}{-\sum_{i=1}^{n} P(i) \ln P(i) - (-\sum_{i=1}^{n} \frac{1}{n} \ln \frac{1}{n})} \tag{5.8}$$

We calculate the network information entropy from 50 constructed gas-water FDCNs. Figure 5.6 shows the variations of network information entropy for

Fig. 5.6 The network
information entropy
distribution in semi-log scale
of 50 gas-water FDCNs with
the increase of gas superficial
velocity

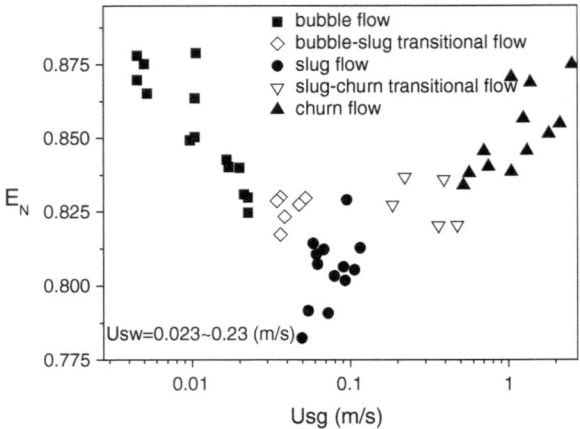

changing flow patterns. We could see that the network information entropy
decreases as the flow pattern evolves from bubble flow to slug flow, but increases
as the flow pattern evolves from slug flow to churn flow.

5.1.4 Nonlinear Dynamics of Gas-Water Two-Phase Flow in FDCN

By mapping the conductance fluctuation signals to corresponding FDCNs, we can
explore the nonlinear dynamics of gas-water two-phase flow from network analysis, which is quantified via a number of topological statistics. Our research team
has discovered that the *Lempel and Ziv complexity* [5, 6], and *approximate entropy*
[7] are sensitive to the flow pattern transition in gas-water two-phase flow. Here we
show in Fig. 5.7, the variations of these two complexity measures with the change
in flow pattern. From Figs. 5.5, 5.6, 5.7, we could see that there are good correlations between complexity measures and the FDCN topological properties (i.e.,
power-law exponent and network information entropy). When the gas superficial
velocity is low, due to the stochastic motion of large numbers of small bubbles, the
dynamics of bubble flow are very complex, corresponding to the large power-law
exponent and network information entropy. In the transition from bubble flow to
slug flow, the dynamics of this transitional flow become relative simple, resulting
in the decrease of the two network properties. Due to the periodic alternating
movements between gas plug and liquid plug, the dynamics of slug flow are quite
simple, which explains the decrease of the two network properties as the flow
pattern evolves from bubble flow to slug flow. When the gas superficial velocity is
high, churn flow, which is composed of discrete gas phase and continuous liquid
phase of high turbulent kinetic energy, gradually appears with the fluctuations.
Because of the turbulence effect, the dynamics of churn flow become more

Fig. 5.7 The complexity measures distributions in semi-log scale with the increase of gas superficial velocity. **a** Lempel and Ziv complexity (using four-symbol coarse graining); **b** Approximate entropy

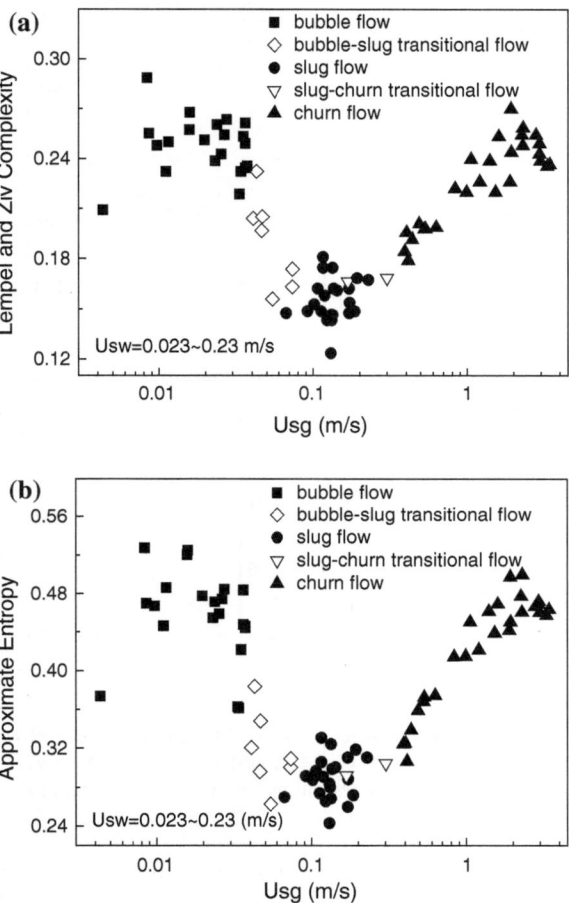

complex than that of slug flow, reflected by the increase of the two network properties as the flow pattern evolves from slug flow to churn flow. Hence, the power-law exponent and network information entropy, which are sensitive to the flow pattern transition, both can characterize the nonlinear dynamics of gas-water two-phase flow.

5.2 Oil-Water Fluid Dynamic Complex Network

We use the method mentioned above to construct the oil-water fluid dynamic complex network [8]. Figure 5.8 indicates that the inclined oil-water FDCN also possesses the scale-free property. Figures 5.9 and 5.10 shows the variations of power-law exponent and network information entropy associated with the change of flow patterns. We see that the power-law exponent and network information

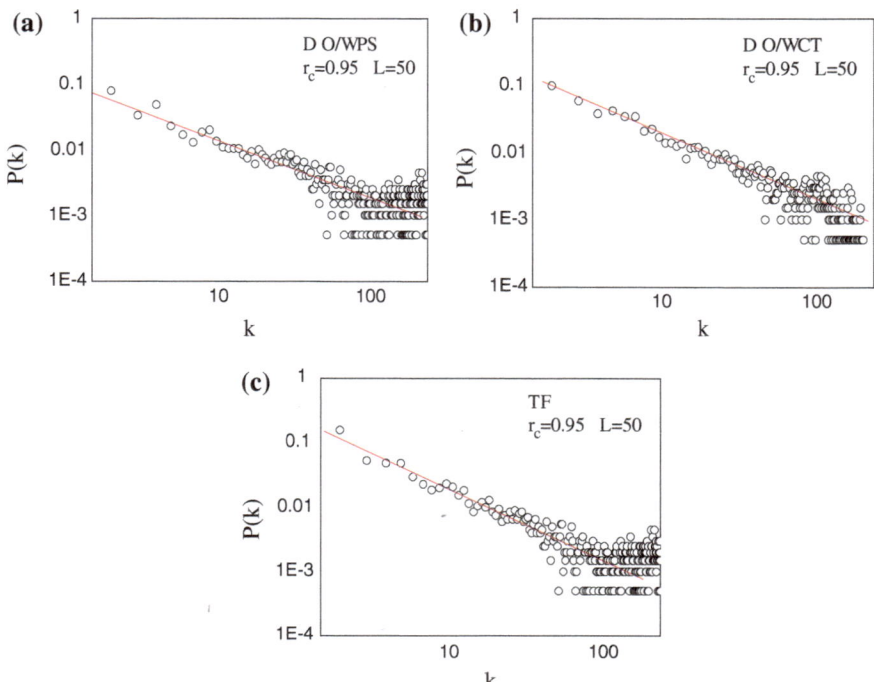

Fig. 5.8 Degree distributions of different types of oil-water FDCNs: **a** D O/W PS flow ($U_{so} = 0.0189$ m/s, $U_{sw} = 0.0061$ m/s); **b** D O/W CT flow ($U_{so} = 0.0816$ m/s, $U_{sw} = 0.0061$ m/s); **c** transitional flow ($U_{so} = 0.1256$ m/s, $U_{sw} = 0.0061$ m/s)

increase as the flow pattern evolves from D O/W PS flow to D O/W CT flow, and further increases as the flow pattern evolves from D O/W CT flow to transitional flow.

Figure 5.11 shows the variations of the two complexity measures associated with the change in flow pattern. From Figs. 5.9, 5.10, 5.11, we could see that there are good correlations between complexity measures and the oil-water FDCN topological properties. The D O/W PS flow can be characterized by a sequence of pseudoslug units, and a pseudoslug unit consists of two major components, an oil pack and a trailing water break. At the front of the oil pack there is normally an oil globule, which is in most cases of elongated shape. The globule is followed by the pack itself, consisting of a large number of droplets with sizes ranging from small to medium. The oil pack is followed by a water break that carries at the top a very few slowly moving droplets left behind by the oil pack. Owing to the periodic alternating movements between oil plug and water plug, the dynamics in the D O/W PS flow is simple, corresponding to the small power-law exponent and network information entropy. When the oil superficial velocity is slightly higher, the D O/W CT flow appears, in which the oil disperses in the continuous water as discrete, well rounded droplets with sizes ranging from mostly small to medium. Because of

Fig. 5.9 Power-law
exponent distribution of 48
oil-water FDCNs with the
increase of oil superficial
velocity

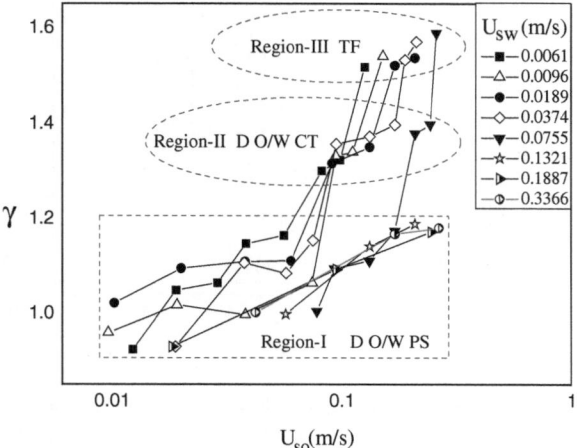

Fig. 5.10 Network
information entropy
distributions of 48 oil-water
FDCNs with the increase of
oil superficial velocity

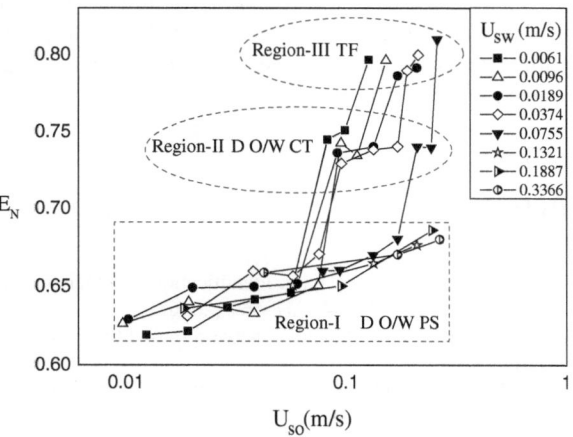

the difference in density, the droplet flow occupies the upper regions of the pipe in an uninterrupted sequence of nearly uniform vertical spread. The outstanding characteristic of this flow pattern consists of local countercurrent flow of water, always at the bottom side of the pipe. And the countercurrent phenomenon is caused by the increase in magnitude of the gravitational component in the direction opposite to the main flow, which partially overcomes the linear momentum of the water phase. Because of the relative complex dynamics in D O/W CT flow, the power-law exponent and the network information entropy increase as the flow pattern evolves from D O/W PS flow to D O/W CT flow. With the further increase of oil superficial velocity, the transitional flow (TF) appears in the region between the water-dominated flow patterns and oil-dominated flow patterns, and the dynamics of this flow becomes more complex than that of D O/W CT flow, indicated by the further increase of the two network statistical characteristics as the

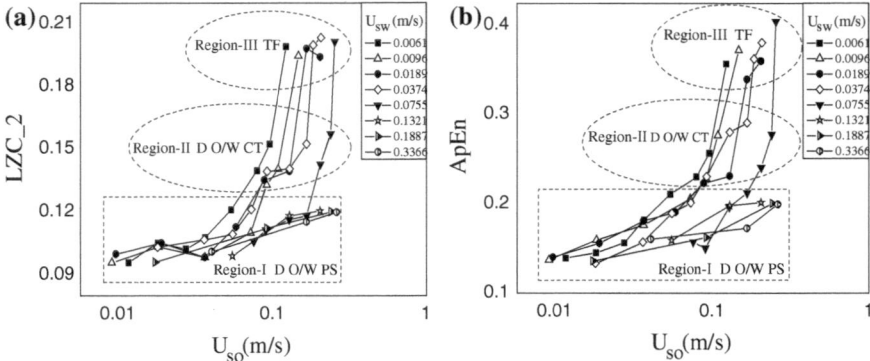

Fig. 5.11 Complexity measure distributions with the increase of oil superficial velocity. Panel **a** shows Lempel and Ziv complexity and panel **b** displays approximate entropies

flow pattern evolves from D O/W CT flow to transitional flow. Hence, the power-law exponent and the network information entropy, which are sensitive to the flow pattern transition, can be used to capture the nonlinear dynamics of inclined oil-water two-phase flow.

References

1. K. Rho, H. Jeong, B. Kahng, Identification of lethal cluster of genes in the yeast transcription network. Physica A **364**, 557–564 (2006)
2. Y. Yang, H.J. Yang, Complex network-based time series analysis. Physica A **387**, 1381–1386 (2008)
3. J.P. Eckmann, S.O. Kamphorst, D. Ruelle, Recurrence plots of dynamical systems. Europhys. Lett. **5**, 973–977 (1987)
4. C.E. Shannon, A mathematical theory of communication. Bell Syst. Tech. J. **27**, 379–423 (1948)
5. A. Lempel, J. Ziv, On the complexity of finite sequences, IEEE Trans. Inf. Theory **22**, 75–81 (1976)
6. F. Kaspar, H.G. Schuster, Easily calculable measure for the complexity of spatiotemporal patterns. Phys. Rev. A **36**(2), 842–848 (1987)
7. S.M. Pincus, Approximate entropy as a measure of system complexity. Proc. Natl. Acad. Sci. U.S.A. **88**, 2297–2301 (1991)
8. Z.K. Gao, N.D. Jin, Complex network analysis in inclined oil-water two-phase flow. Chin. Phys. B **18**(12), 5249–5258 (2009)

Chapter 6
Gas-Water Fluid Structure Complex Network

In general, the traditional nonlinear time series analysis methods (chaotic attractor morphology, complexity measures and chaotic recurrence plot) cannot effectively reveal the complex fluid structure of two-phase flow. Though the network construction algorithm mentioned in Chap. 5 can be applied to investigate the fluid dynamics of two-phase flow to a certain extent, it may be not available to study the fluid structure of two-phase flow. As an alternative, the advantage of utilizing a phase space reconstruction is that if the embedding is chosen appropriately, the topological distribution of the set of vector points in phase space will accurately reflect the underlying dynamics of the original system. Therefore, the network inferred from that phase space reconstruction can directly link to the evolution operator of the underlying dynamical system.

Recently a new framework has been established to construct Phase-Space Complex Network (PSCN) from signal time series [1]. In this chapter, we introduce our proposed phase-space complex network. In addition, we bridge different aspects of the dynamics of the time series with the topological indices of the network to demonstrate how such topological properties can be used to distinguish different dynamical regimes.

6.1 Phase-Space Complex Network from Time Series

Our method [1, 2] to construct a phase-space complex network from a time series can be described as follows. Given a time series $z(it)$, $i = 1, 2..., M$, where t is the sampling interval and M is the sample size, we construct a sequence of phase-space vectors according to the standard delay coordinate embedding method [3–5],

$$\begin{aligned}
\vec{X_k} &= \{x_k(1), x_k(2), \ldots, x_k(m)\} \\
&= \{z(kt), z(kt + \tau), \ldots, z(kt + (m - 1)\tau)\}
\end{aligned} \tag{6.1}$$

where τ is the delay time, m is the embedding dimension, $k = 1, 2, \ldots, N$, and $N = M - (m - 1) \tau/t$ is the total number of vector points in the reconstructed phase space. There are various empirical criteria for choosing the delay time τ, and

Z.-K. Gao et al., *Nonlinear Analysis of Gas-Water/Oil-Water*
Two-Phase Flow in Complex Networks, SpringerBriefs on Multiphase Flow,
DOI: 10.1007/978-3-642-38373-1_6, © The Author(s) 2014

we have used a correlation-integral-based method, e.g., C–C method, (see, e.g., reference [6]) for this purpose. C–C method is the frequent method that can be used to determine the delay time τ. The general idea of C–C method is that, we can subdivide a single time series $\{z_1, z_2, \ldots, z_M\}$ into t disjoint time series, where t is the index lag, and then calculate $S(m, M, r, t)$ from these time series as follows:

$$S(m, M, r, t) = \frac{1}{t} \sum_{s=1}^{t} \left[C_s(m, M/t, r, t) - C_s^m(1, M/t, r, t) \right] \tag{6.2}$$

$$\Delta S(m, t) = \max \left(S(m, r_j, t, M) \right) - \min \left(S(m, r_j, t, M) \right) \tag{6.3}$$

where

$$C(m, M, r, \tau) = \frac{2}{N(N-1)} \sum_{1 \leq i < j \leq N} \theta \left(r - |x_i - x_j| \right), \quad N = M - (m-1)\tau, \, r > 0,$$

and $\theta(a)$ is the Heaviside function. The general range of m, r is $2 \leq m \leq 5, \frac{\sigma}{2} \leq r \leq 2\sigma$, σ, is the standard deviation of data sets.

$$\overline{S}(t) = \frac{1}{16} \sum_{m=2}^{5} \sum_{j=1}^{4} S(m, r_j, t) \tag{6.4}$$

$$\Delta \overline{S}(t) = \frac{1}{4} \sum_{m=2}^{5} \Delta S(m, t) \tag{6.5}$$

The corresponding time of the first local minimum of $\Delta \overline{S}(t)$ is the optimal delay time τ. More details about C–C method can be found in reference [6]. For determining the embedding dimension, there exists a rigorous mathematical criterion [4, 7]. For noisy time series, it is convenient to use some heuristic criterion such as the one based on distinguishing false nearest neighbors (FNNs) [5] in the reconstructed phase space. The FNN method can generate a minimum embedding dimension m that ensures unfolding of orbits in the phase space, so that there are no false nearest neighbors for every orbit point. In contrast, if the embedding dimension is less than m, orbits in the phase space cannot be fully unfolded. After m is determined, we employ the correlation-integral-based algorithm [6] to determine τ. We note that if τ is too small, the reconstructed attractor can be compressed along the identity line and, if τ is too large, trajectories on the attractor may become disconnected. Our choice of τ avoids these undesirable situations.

To construct a network, we regard each vector point as a node and use the phase-space distance to determine the edges. Given two vector points $\vec{X_i}$ and $\vec{X_j}$, the phase-space distance is defined to be

$$d_{ij} = \sum_{n=1}^{m} \left\| X_i(n) - X_j(n) \right\| \tag{6.6}$$

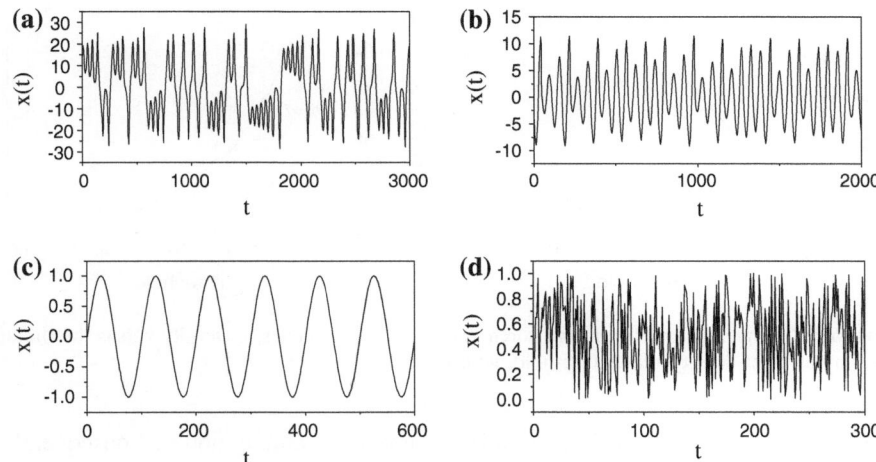

Fig. 6.1 Four different scalar time series that are used to construct complex networks. **a** Chaotic time series from Lorenz system ($\sigma = 16$, $\rho = 45.92$, $\beta = 4$); **b** Chaotic time series from Rössler system ($a = 0.2$, $b = 0.2$, $c = 5.7$); **c** Periodic time series; **d** White noise time series

where $X_i(n) = z(i + (n - 1)\tau)$ and $X_j(n) = z(j + (n - 1)\tau)$ is the nth element of $\vec{X_i}$ and $\vec{X_j}$, m and τ is the embedding dimension and delay time, respectively. This generates, for all nodes (vector points) in the network, a distance matrix $D = (d_{ij})$. Choosing a critical threshold r_c, the distance matrix $D = (d_{ij})$ can be converted into adjacency matrix $A = (A_{ij})$, the rules of which read: $A_{ij} = 1$ if $d_{ij} \leq r_c$ and $A_{ij} = 0$ if $d_{ij} > r_c$. Thus, an edge connecting node i and j exists if $A_{ij} = 1$, while there is no edge between i and j if $A_{ij} = 0$. The topology of the reconstructed PSCN is determined entirely by A.

An appropriate threshold therefore should be chosen to fully preserve the main property of the network, but too large values may obscure or conceal the local property by over-connecting the nodes. We employ the network density [8] to study the threshold, which is defined as the number of edges divided by the largest number of edges possible. In order to show how to select a suitable threshold, a chaotic time series from Lorenz system (see Fig. 6.1a) is studied by this method, and Fig. 6.2 shows the density of the constructed network versus the threshold r. As shown in Fig. 6.2b, the increase of degree reaches the maximum rate at about $r_c = 7.6$, which we set to be the critical threshold. The reason for setting the threshold r at this critical point is as follows. It can be imagined that the node degrees increase more rapidly as the threshold changes within the cluster radius due to the adjacency of the nodes inside. For a network from a chaotic system that has many clusters differing in sizes, it can be expected that the maximum increase rate of the number of edges arises as the threshold approaches the critical point r_c, i.e., the "mean radius" of all the clusters. The network obtained at r_c will maintain the clustering property, and thresholds beyond this value will result in a much

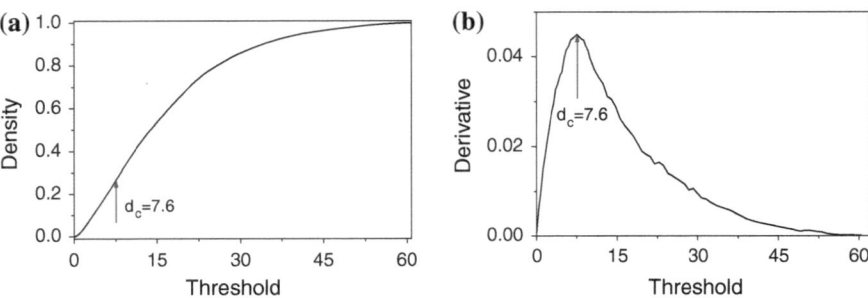

Fig. 6.2 Complex network from the Lorenz system with 600 nodes. **a** Density versus threshold; **b** Derivative of density versus threshold, with $r_c = 7.6$

slower edge increase, causing redundant connections among nodes. Consequently we choose r_c to construct the network. Furthermore, it should be noted that r_c will depend on the size of the network (the length of the time series).

We apply the method presented in this paper to generate networks from Lorenz system, Rössler system, periodic time series and white noise time series, respectively (details of the four different time series can be seen in Fig. 6.1). We find that the networks constructed from different kind of time series demonstrate fundamentally difference. It is noteworthy that the structure of constructed network, i.e., Fig. 6.3, is drawn by the software "Netdraw", and the algorithm used is the Kamada-Kawai spring embedding algorithm [9]. The general idea of Kamada-Kawai spring embedding algorithm is as follows: First, calculates the energy for each node. It then loops over all the nodes to find the one with the highest energy, and begins iterating the Newton-Raphson stage to compute new positions for the node until its energy is below epsilon. At this point, it again looks for the node with the highest energy and begins moving it (more details see reference [9]). This process continues until there is no node with energy above epsilon. As shown in Fig. 6.3a, b, the network structure, corresponding to chaotic time series, has nodes congregating at different locations, and some regions are highly clustered with nodes and others are rather sparse. The network generated from periodic time series in Fig. 6.3c is apparently a regular network. In contrast, the network from the white noise time series looks like a random network, as shown in Fig. 6.3d.

Before further analysis, we here give an intuitive description of the loops-layout in phase space for chaotic systems in terms of the unstable periodic orbits (UPOs). Unstable periodic orbits embedded in the chaotic attractor are fundamental to the understanding of the chaotic dynamics [10–17]. For a chaotic attractor, its trajectory will typically switch or hop among different UPOs. Specifically, the trajectory will approach an unstable periodic orbit along its stable manifold. This approach can last for several loops during which the orbit remains close to the UPO. Eventually, the orbit is ejected along the unstable manifold and proceeds until it is captured by the stable manifold of another UPO. A UPO of order n contains n loops lying in different locations in phase space. Each loop that

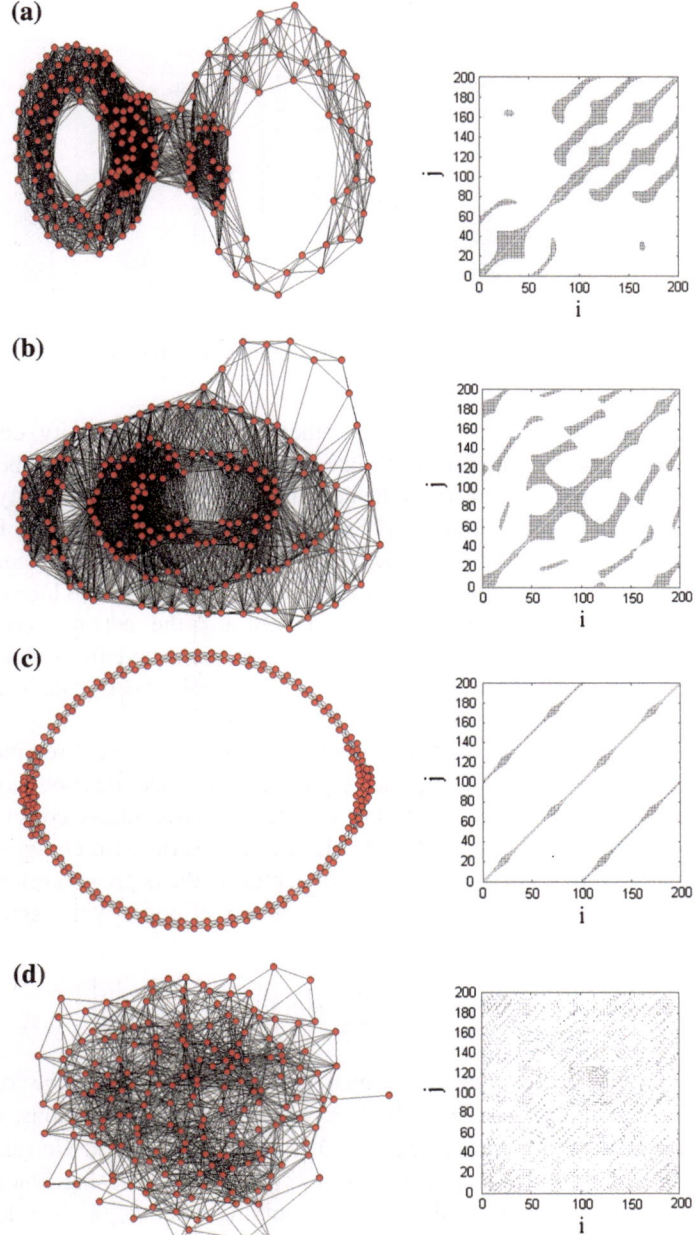

Fig. 6.3 Complex network of 200 nodes from **a** Lorenz system with $m = 3$, $\tau = 10$, $r_c = 11.2$; **b** Rössler system with $m = 3$, $\tau = 13$, $r_c = 6.2$; **c** periodic time series with $m = 4$, $\tau = 3$, $r_c = 0.15$ and **d** white noise time series with $m = 4, \tau = 5, r_c = 0.41$. The structure of network drawn by the software "Ucinet" and "Netdraw" are shown in the *left panel*, and the corresponding network connectivity (adjacency) matrix are shown in the *right panel*

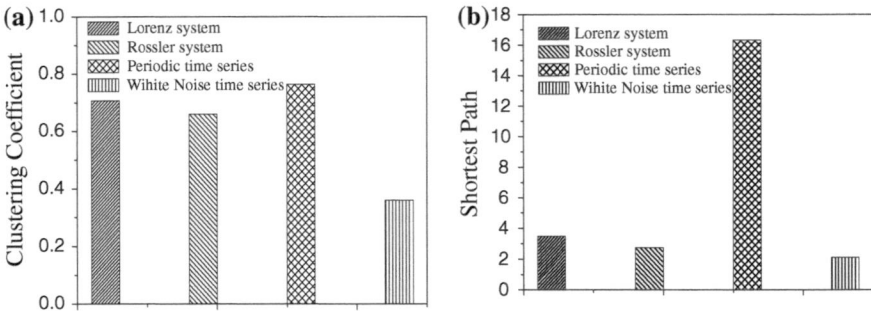

Fig. 6.4 Small-world characteristics of the networks from chaotic time series

belongs to a certain UPO-n has many other loops in its vicinity due to the attraction of the stable manifold associated with the UPO-n. It then becomes a center of a cluster and the density of the neighbors is related to its stability decided by the vector field along its trajectory. Since the stability of each center loop may vary, we can see sparse as well as dense regions in the structure of constructed complex network, as shown in Fig. 6.3a, b. Due to the fact that the loops in phase space are spatially clustered around the UPOs and the network connections determined by phase space, the complex network generated from chaotic time series shows the small-world characteristics, i.e., small shortest path and large clustering coefficient, which can be seen in Fig. 6.4.

We now demonstrate how to discriminate different dynamical regimes of the time series through investigating the degree correlations, Pearson coefficient, betweenness distribution and clustering coefficient-betweenness correlations of two distinct complex networks from the chaotic time series (Lorenz system) and white noise time series. An important way of capturing the degree correlations is to examine the average degree of the nearest neighbors of nodes with degree k [18], which is defined as:

$$k_{nn}(k) = \sum_{k'} k' p(k'|k) \tag{6.7}$$

where $p(k'|k)$ denotes the conditional probability that an edge of degree k connects a node with degree k'. If there are no degree correlations, Eq. (6.7) gives $k_{nn}(k) = <k^2> / <k>$, i.e., $k_{nn}(k)$ is independent of k. Correlated networks are classified as assortative mixing if $k_{nn}(k)$ is an increasing function of k, whereas they are referred to as disassortative mixing when $k_{nn}(k)$ is a decreasing function of k. In other words, in assortative networks the nodes tend to connect to their connectivity peers, while in disassortative networks nodes with low degrees are more likely connected with highly connected ones. As shown in Fig. 6.5, $k_{nn}(k)$ increases with k for chaotic Lorenz system, while decreases with k for noisy time series. In order to quantify such a correlation, we calculate the Pearson coefficient [19, 20] which is defined as follows. Let $e_{kl}(= e_{lk})$ be the joint

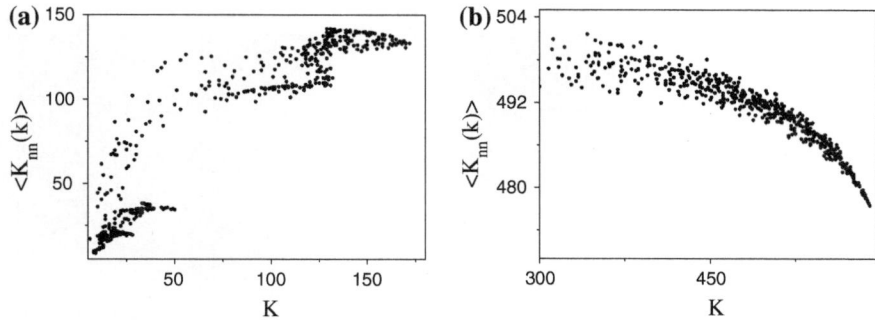

Fig. 6.5 The regular plots of the nearest neighbor average connectivity of nodes with respect to connectivity of the networks containing 600 nodes from **a** Lorenz system with $m = 3$, $\tau = 23$, $r_c = 10$ and **b** white noise time series with $m = 4$, $\tau = 3$, $r_c = 0.8$

probability distribution for an edge to be associated with a node with connectivity k at one end and a node with connectivity l at the other. Its marginal, $q_k = \sum_l e_{kl}$, obeys the normalization condition, $\sum_k q_k = 1$. Then, the Pearson correlation coefficient is given by

$$r = \frac{1}{\sigma_q^2} \sum_{k,l} kl(e_{kl} - q_k q_l) \tag{6.8}$$

where $\sigma_q^2 = \sum_k k^2 q_k - \left(\sum_k k q_k\right)^2$ is the variance of $q_k \cdot r \in [-1, 1]$, and r is positive (negative) for assortative (disassortative) mixing. The Pearson coefficient of the network containing 600 nodes from chaotic Lorenz system and noisy time series is 0.1009 and −0.0037, respectively. Consistently with the one obtained from the analysis of the nearest neighbor average connectivity, the Pearson correlation coefficient is positive, confirming that the network from chaotic time series is assortative mixing. On the other hand, the network from noisy time series is of neutral assortative mixing. We can explain why the network corresponding to chaotic time series possesses the property of assortative mixing in terms of UPOs. For a complex network generated from chaotic time series, there are multiple clusters and most nodes connected to each other within the same cluster have a roughly similar number of connected neighbors or degrees. Because of the different stability of the center loop of UPOs, the common degree shared by the nodes from one cluster may differ from that of another. Due to the fact that the nodes within same cluster usually have similar degrees and such nodes are always interconnected to each other, chaotic time series possesses the property of assortative mixing.

Another way of capturing the clustering property of chaotic time series is the sorted adjacent matrix proposed in this paper. That is, for a given complex network, we first sort nodes in ascending order of the degree. If two or more than two nodes have the same degree, we then arrange them in ascending order of the sum of neighbor degree, which is the sum of degree of a node's neighbors. Finally, the

(a)   **(b)**

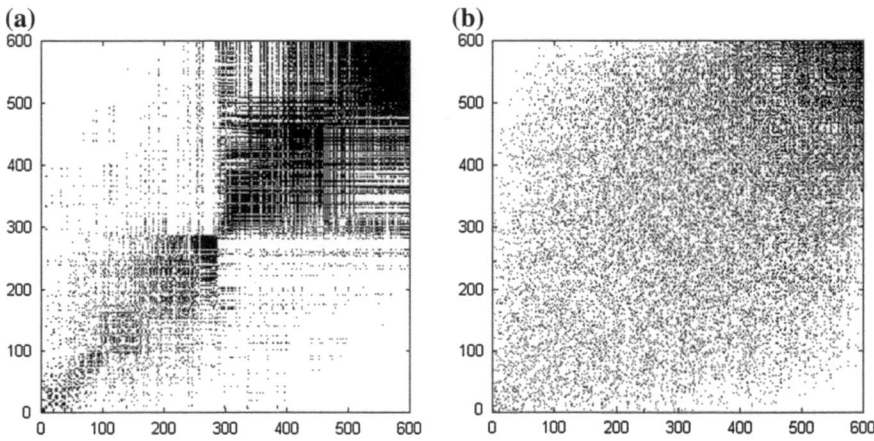

Fig. 6.6 The bitmap of sorted adjacent matrix of network containing 600 nodes from **a** Lorenz system with $m = 3$, $\tau = 23$, $r_c = 10$ and **b** white noise time series with $m = 4$, $\tau = 3$, $r_c = 0.8$

above node sorting rules produce a sorted adjacent matrix, which can be represented as a black-and-white bitmap, i.e., if nodes i and j are connected, a black pixel is placed at the coordinate of (i, j); otherwise a white pixel is placed there. Figure 6.6 depicts the bitmap of sorted adjacent matrix of two networks from different time series, i.e., chaotic series and noisy series. In particular, multiple black square blocks could be clearly found along the principal diagonal in the sorted adjacent matrix bitmap of the network from chaotic time series, which implies the fact that there are multiple clusters in the network and the nodes with similar degrees are interconnected to each other (i.e., assortative mixing). On the contrary, no obvious black square block could be detected along the principal diagonal in the sorted adjacent matrix bitmap corresponding to the noisy time series.

From the above analysis, we can see that the clustering property of chaotic time series can be well characterized by the degree correlations, Pearson coefficient as well as the sorted adjacent matrix, and for networks from the noisy time series, no assortative mixing is found at different thresholds. Assortative mixing provides the information about the interactions of all node pairs. In addition, we may also concern how important, or how central a single node is in a network. A concept that fulfills this requirement is betweenness, or betweenness centrality. Together with the degree and the closeness of a node, the betweenness is one of the standard measures of node centrality, originally introduced to quantify the importance of an individual in a social network [21–23]. More precisely, the betweenness b_i of a node i, referred to also as load, is defined as:

$$b_i = \sum_{j,k \in N, j \neq k} \frac{n_{jk}(i)}{n_{jk}} \tag{6.9}$$

where n_{jk} is the number of shortest paths connecting node j and k, while $n_{jk}(i)$ is the number of shortest paths connecting j and k and passing through i. Through

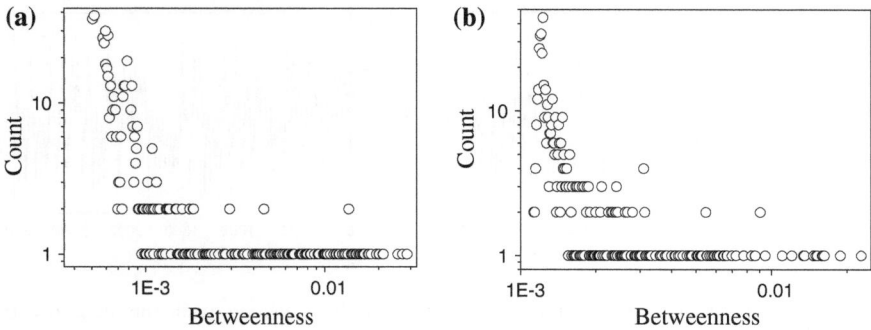

Fig. 6.7 Betweenness distribution in log-log scale for complex network of 600 nodes from **a** Lorenz system with $m = 3$, $\tau = 23$, $r_c = 10$, **b** white noise time series with $m = 4$, $\tau = 3$, $r_c = 0.8$

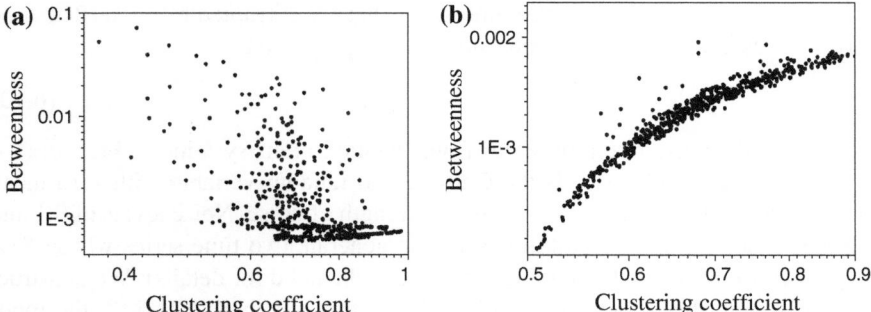

Fig. 6.8 Clustering coefficient-betweenness correlations in log–log scale for complex network of 600 nodes from **a** Lorenz system with $m = 3$, $\tau = 23$, $r_c = 10$, **b** white noise time series with $m = 4$, $\tau = 3$, $r_c = 0.8$

exploring the betweenness distribution and clustering coefficient-betweenness correlations, we not only show in Fig. 6.7 that the overall betweenness of the network constructed from chaotic time series is different from that of the network generated from noisy time series, but also demonstrate in Fig. 6.8 that a large number of nodes with small clustering coefficient have high betweenness in the network from chaotic time series, in contrast to the high betweenness only for the nodes with large clustering coefficient in the network from noisy time series. The distinct distributions imply that the vector points are structured with different mechanisms in the phase space. We find that for the chaotic time series, the number of nodes with high betweenness greatly exceeds that from the noisy time series. This is essentially a reflection of the clustering property associated with the UPOs embedded in the chaotic attractor. The high betweenness nodes correspond to the nodes in between adjacent clusters that act as bridges, and such nodes usually have small clustering coefficient compared with the nodes within the

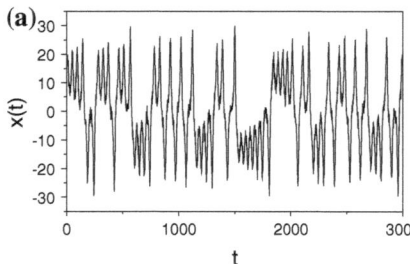 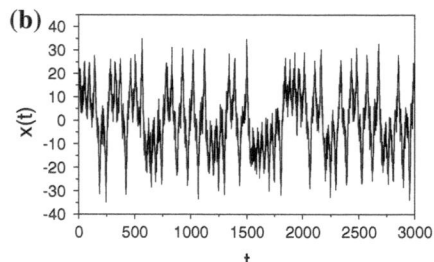

Fig. 6.9 Two different chaotic time series corrupted by Gaussian white noise. **a** $\eta = 0.1$ (*SNR* = 10 dB); **b** $\eta = 0.3$ (*SNR* = 5 dB)

clusters. Since a chaotic attractor contains infinitely UPOs, there will be many clusters in the corresponding network, which implies the existence of numerous high betweenness nodes.

Since most real data contains noise, we now discuss the anti-noise ability of the method by analyzing two chaotic time series that are corrupted by noise. The time series we used are generated from the following equation

$$z_i = L_i + \eta \sigma \varepsilon_i \qquad (6.10)$$

where L_i is the noise-free time series from chaotic Lorenz system(see Fig. 6.1a), σ is the standard deviation, ε_i is the Gaussian iid random variable with zero mean and a standard deviation of 1, and η is the strength of noise. Noise levels of 0.1 and 0.3 were added to the Lorenz time series to generate two time series whose *SNR* (Signal Noise Ratio) is 10 and 5 dB (see Fig. 6.9a and d for details). We construct networks from these two time series and find that, when $SNR > 5dB$, the topological structure of constructed network attractor is very similar to the one that constructed from deterministic chaotic Lorenz system, as shown in Figs. 6.3a and 6.10a; when the noise intensity becomes very large, i.e., $SNR \leq 5dB$, because of the strong influence of Gaussian white noise, the corresponding topological structure distorts to a certain extent, as shown in Fig. 6.10b. To gain insight into an important question of whether the topological properties still can be used to distinguish chaotic time series corrupted by Gaussian white noise from white noise time series, we investigate the degree correlations, betweenness distribution and clustering coefficient-betweenness correlations of the two series presented in Fig. 6.9. The increase of $k_{nn}(k)$ with k in Fig. 6.11 and the multiple black square blocks along principal diagonal in Fig. 6.12 both indicate the fact that the chaotic time series in the presence of Gaussian white noise also possess the property of assortative mixing. As shown in Figs. 6.13 and 6.14, the clustering property associated with the UPOs of the chaotic time series corrupted by Gaussian white noise also can be reflected by the betweenness distribution and clustering coefficient-betweenness correlations. From the above analysis, we claim that our method can also be applied to distinguish chaotic time series with Gaussian white noise. Considering the fact that SNR of most real data is greater than 5 dB, we can infer that the phase-space complex network has good anti-noise ability.

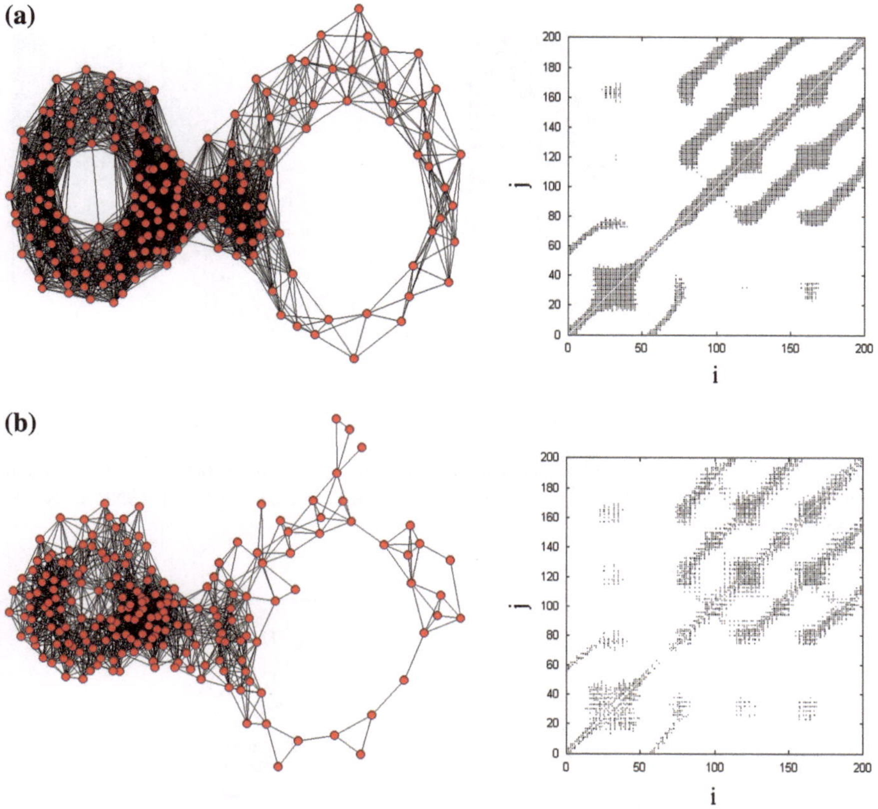

(a)

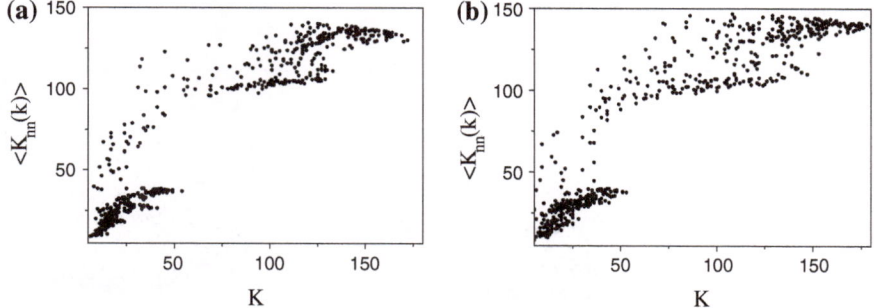

(b)

Fig. 6.10 Complex network of 200 nodes from chaotic time series corrupted by Gaussian white noise. **a** Noise intensity $\eta = 0.1$ (SNR = 10 dB), **b** Noise intensity $\eta = 0.3$ (SNR = 5 dB). The structure of network drawn by the software "Ucinet" and "Netdraw" are shown in the *left panel*, and the corresponding network connectivity (adjacency) matrix are shown in the *right panel*

(a)

(b)

Fig. 6.11 The regular plots of the nearest neighbor average connectivity of nodes with respect to connectivity of the networks containing 600 nodes from chaotic time series corrupted by Gaussian white noise. **a** Noise intensity $\eta = 0.1$ (SNR = 10 dB), **b** Noise intensity $\eta = 0.3$ (SNR = 5 dB)

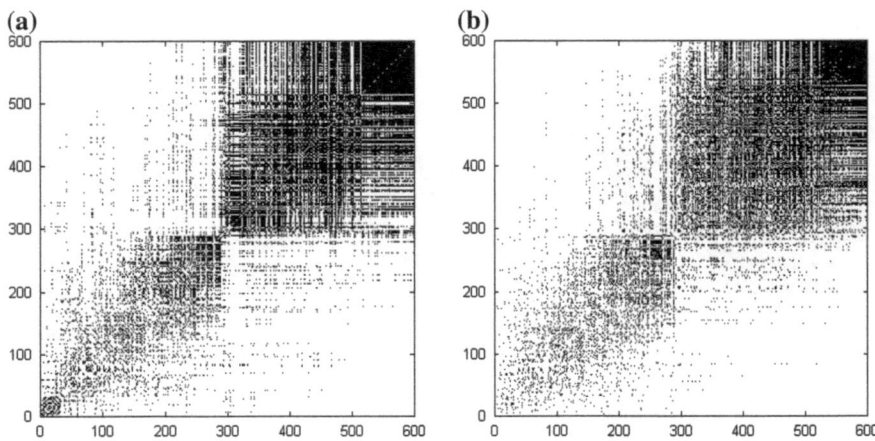

Fig. 6.12 The bitmap of sorted adjacent matrix of network containing 600 nodes from chaotic time series corrupted by Gaussian white noise. **a** Noise intensity $\eta = 0.1$ (SNR = 10 dB), **b** Noise intensity $\eta = 0.3$ (SNR = 5 dB)

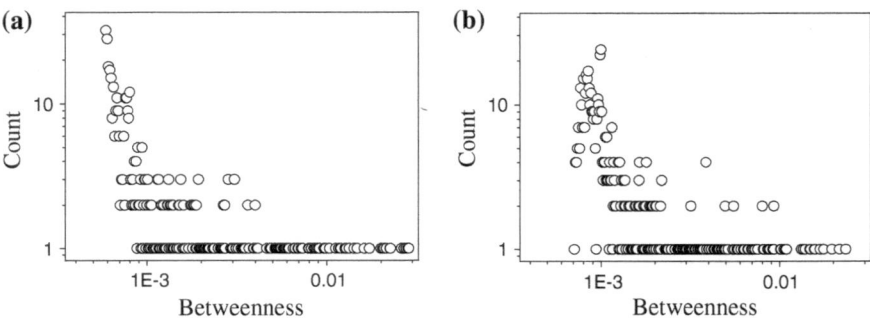

Fig. 6.13 Betweenness distribution in log–log scale for complex network of 600 nodes from chaotic time series corrupted by Gaussian white noise. **a** Noise intensity $\eta = 0.1$ (SNR = 10 dB), **b** Noise intensity $\eta = 0.3$ (SNR = 5 dB)

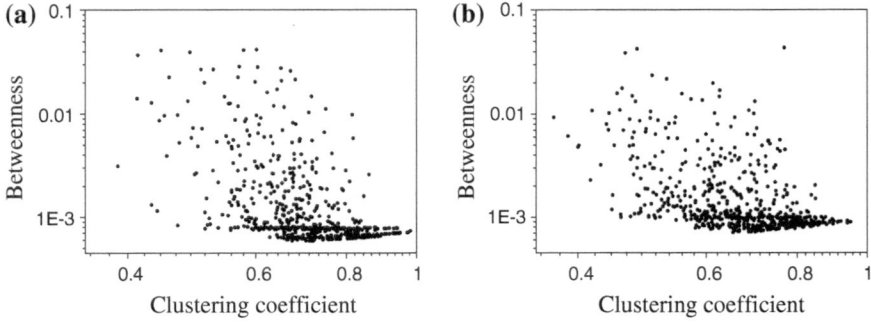

Fig. 6.14 Clustering coefficient-betweenness correlations in log–log scale for complex network of 600 nodes from chaotic time series corrupted by Gaussian white noise. **a** Noise intensity $\eta = 0.1$ (SNR = 10 dB), **b** Noise intensity $\eta = 0.3$ (SNR = 5 dB)

Fig. 6.15 The structure of FSCN containing 200 nodes from **a** Bubble flow ($Q_g = 0.2$ m^3/h, $Q_w = 2.0$ m^3/h) with $r_c = 0.012$, **b** Slug flow ($Q_g = 4.1$ m^3/h, $Q_w = 6.0$ m^3/h) with $r_c = 0.16$, **c** Churn flow ($Q_g = 139.0$ m^3/h, $Q_w = 2.0$ m^3/h) with $r_c = 0.11$

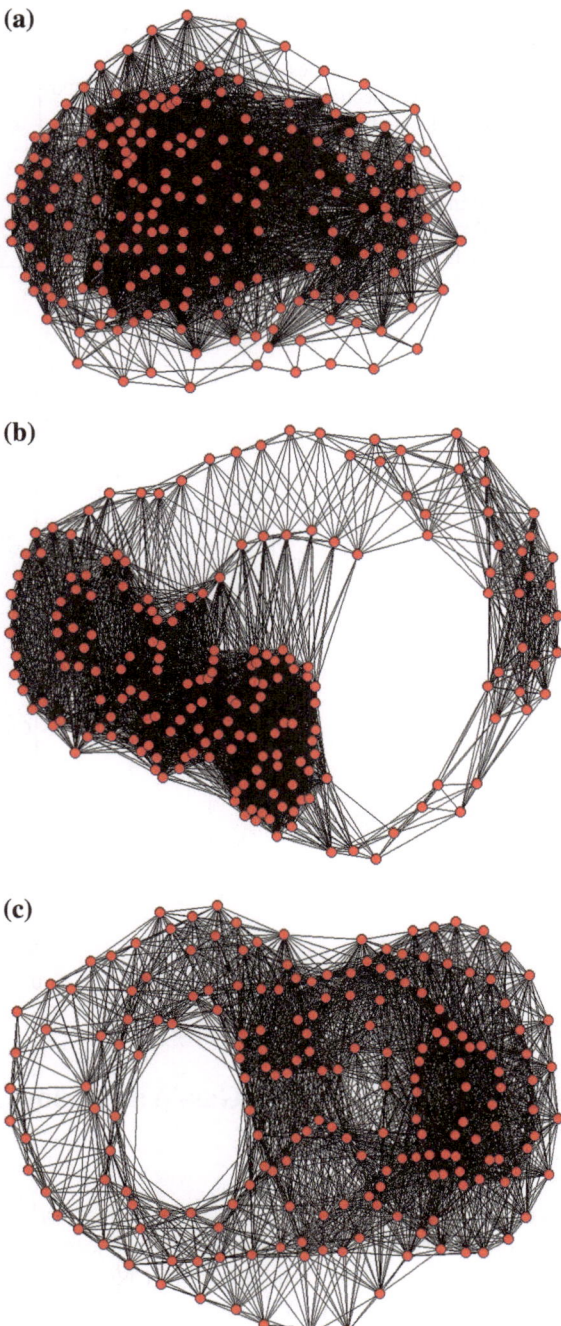

Fig. 6.16 The regular plots of the nearest neighbor average connectivity of nodes with respect to connectivity of the networks containing 2000 nodes from **a** Bubble flow ($Q_g = 0.2$ m^3/h, $Q_w = 2.0$ m^3/h) with $r_c = 0.031$, **b** Slug flow ($Q_g = 4.1$ m^3/h, $Q_w = 6.0$ m^3/h) with $r_c = 0.35$, **c** Churn flow ($Q_g = 139.0$ m^3/h, $Q_w = 2.0$ m^3/h) with $r_c = 0.26$

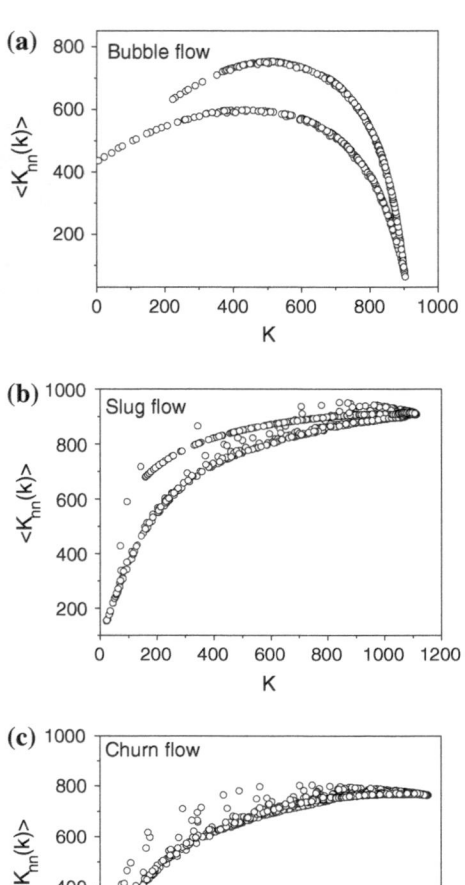

6.2 Fluid Structure of Gas-Water Flow in Fluid Structure Complex Network

We construct Fluid Structure Complex Network (FSCN) using the phase-space complex network method to carry out more investigations on their assortative mixing property. The structure of FSCNs generated from three typical flow patterns are shown in Fig. 6.15.

We will demonstrate how the statistic of FSCN can be used to reveal the fluid structure of gas-water two-phase flow. In the bubble flow, the gas phase is approximately uniformly distributed in the form of discrete small bubbles in a continuum of liquid phase, and nearly no bubble coalescence can be observed. While in the slug flow, due to the bubble coalescence, many large bubbles which have a diameter almost equal to the pipe diameter appear. Bubble coalescence, which has an important impact on the fluid structure of gas-water two-phase flow, corresponds to many unstable periodic orbits (UPOs) in the reconstructed phase space. For a FSCN from slug flow, there are multiple clusters which are caused by the bubble coalescence, and most nodes connected to each other within the same cluster have a roughly similar number of neighbors or node degrees. The common degree shared by the nodes from one cluster may differ from that of another because of the different stability of the center loop of UPOs. This has led to the fact that the FSCN from slug flow possesses the property of strong assortative mixing (Details see Fig. 6.16b), while no assortative mixing can be found in the FSCN from bubble flow (see Fig. 6.16a). Churn flow can be treated as an irregular, chaotic and disordered slug flow. Since bubble coalescence and bubble collapse both exist in the fluid structure of churn flow, the FSCN from churn flow is assortative mixing. But compared to the FSCN from slug flow, the FSCN from churn flow shows weaker assortative mixing, which may be caused by the bubble collapse (Details see Fig. 6.16c).

Therefore, the fluid structure of gas-water flow has been uncovered in terms of the topological indices of FSCN, and in particular, the assortative mixing property of FSCN can effectively reveal the gas-water fluid structure to some extent.

References

1. Z.K. Gao, N.D. Jin, Complex network from time series based on phase space reconstruction. Chaos **19**, 033137 (2009)
2. Z.K. Gao, N.D. Jin, W.X. Wang, Y.C. Lai, Motif distributions in phase-space networks for characterizing experimental two-phase flow patterns with chaotic features. Phys. Rev. E **82**(2), 016210 (2010)
3. N.H. Packard, J.P. Crutchfield, J.D. Farmer, Geometry from a time series. Phys. Rev. Lett. **45**(9), 712–716 (1980)
4. T. Sauer, J.A. Yorke, M. Casdagli, Embedology. J. Stat. Phys. **65**, 579–616 (1991)
5. M.B. Kennel, R. Brown, H.D.I. Abarbanel, Determining embedding dimension for phase-space reconstruction using a geometrical construction. Phys. Rev. A **45**(6), 3403–3411 (1992)
6. H.S. Kim, R. Eykholt, J.D. Salas, Nonlinear dynamics, delay times, and embedding windows. Physica D **127**, 48–60 (1999)
7. F. Takens, *Dynamical Systems and Turbulence, Lecture Notes in Mathematics*, vol 898 (Springer, New York, 1981), pp. 366–381
8. R.V. Donner, J. Heitzig, J.F. Donges, Y. Zou, N. Marwan, J. Kurths, The geometry of chaotic dynamics—a complex network perspective. Eur. Phys. J. B **84**, 653–672 (2011)
9. T. Kamada, S. Kawai, An algorithm for drawing general undirected graphs. Inform. Process. Lett. **31**(1), 7–15 (1989)
10. D. Auerbach, P. Cvitanovic, J.P. Eckmann, G. Gunaratne, I. Procaccia, Exploring chaotic motion through periodic orbits. Phys. Rev. Lett. **58**(23), 2387–2389 (1987)
11. B. Hunt, E. Ott, Optimal periodic orbits of chaotic systems. Phys. Rev. Lett. **76**(13), 2254–2257 (2004)

12. Y.C. Lai, Y. Nagai, C. Grebogi, Characterization of the natural measure by unstable periodic orbits in chaotic attractors. Phys. Rev. Lett. **79**, 649–652 (1997)
13. M. Dhamala, Y.C. Lai, Unstable periodic orbits and the natural measure of nonhyperbolic chaotic saddles. Phys. Rev. E **60**, 6176–6179 (1999)
14. R.L. Davidchack, Y.C. Lai, A. Klebanoff, E.M. Bollt, Toward complete detection of unstable periodic orbits in chaotic systems. Phys. Lett. A **287**, 99–104 (2001)
15. M. Dhamala, Y.C. Lai, The natural measure of nonattracting chaotic sets and its representation by unstable periodic orbits. Int. J. Bifurcat. Chaos **12**, 2991–3006 (2002)
16. A. Akaishi, A. Shudo, Accumulation of unstable periodic orbits and the stickiness in the two-dimensional piecewise linear map. Phys. Rev. E **80**(6), 026211 (2009)
17. S.H. Wu, J.H. Hao, H.B. Xu, Controlling chaos to unstable periodic orbits and equilibrium state solutions for the coupled dynamos system. Chin. Phys. B **19**(2), 020509 (2010)
18. R. Pastor-Satorras, A. Vazquez, A. Vespignani, Dynamical and correlation properties of the internet. Phys. Rev. Lett. **87**(25), 258701 (2001)
19. M.E.J. Newman, Assortative mixing in networks. Phys. Rev. Lett. **89**(20), 208701 (2002)
20. M.E.J. Newman, Mixing patterns in networks. Phys. Rev. E **67**(2), 026126 (2003)
21. C.L. Freeman, A set of measures of centrality based on betweenness. Sociometry **40**, 35–41 (1977)
22. C.L. Freeman, Centrality in social networks: conceptual clarification. Social networks **1**, 215–239 (1979)
23. S. Wasserman, K. Faust, *Social Networks Analysis* (Cambridge University Press, Cambridge, 1994)

Chapter 7
Oil-Water Fluid Structure Complex Network

Due to the interplay among many complex factors such as liquid-liquid phase interfacial interaction and the existence of a gravitational component normal to the flow direction, an inclined oil-water flow exhibits highly irregular, random, and unsteady flow structure as compared with a vertical two-phase flow. Moreover, since there does not exist obvious bubble coalescence in inclined oil-water two-phase flow, the assortative mixing property of FSCN is always ineffective for uncovering the complex oil-water fluid structure. Arm to the inclined oil-water two-phase flow, we have developed a unique approach for investigate its fluid structure [1]. That is, we exploit the concept of network motifs [2, 3] to characterize inclined oil-water two-phase flow. Network motifs have been found to be fundamental to gene regulatory networks in systems biology [2–6] and are also useful for characterizing networks from other disciplines [7–12]. We develop a different network construction approach and apply it to experimental oil-water two-phase flows to investigate different flow structures in terms of motif distributions. Specifically, given a set of conductance fluctuating signals from some inclined oil-water two-phase flow experiment, our first step is to construct oil-water fluid structure complex network using the PSCN construction method mentioned in Chap. 6. We then search for possible motifs from the constructed network and calculate their distributions. Our main result is that motif distributions do exist in the constructed networks and, strikingly, they tend to be highly heterogeneous, a feature that has been found to be common for PSCNs constructed from low-dimensional deterministic chaotic systems. The motif distribution can thus faithfully represent the distinct dynamical states of the two-phase flow. For example, when a transition in the flow pattern occurs, a characteristic change in the motif distribution arises. The results suggest that motif distribution can potentially be a powerful tool for revealing the nonlinear dynamics of oil-water two-phase flows.

We using the PSCN construction method construct network. It should be pointed out that the method of choosing the threshold presented in Chap. 6 is effective for small-size networks. For large networks, the method can result in unrealistic values of the threshold, leading possibly to loss of information about the local phase-space or motif structure. Take, for example, networks

Z.-K. Gao et al., *Nonlinear Analysis of Gas-Water/Oil-Water*
Two-Phase Flow in Complex Networks, SpringerBriefs on Multiphase Flow,
DOI: 10.1007/978-3-642-38373-1_7, © The Author(s) 2014

reconstructed from chaotic systems. We are interested in exploring the interplay between network motifs and the fundamental building blocks of chaotic set. Thus, we choose r_c to be about 18–20 % of the root-mean-square (rms) value of the measured signal to investigate the local property of large phase-space complex network. To illustrate the structure of large PSCNs from typical chaotic systems, we consider three classical examples: (1) chaotic Lorenz system given by

$$\begin{cases} dx/dt = 16(y - x) \\ dy/dt = x(45.92 - z) - y \\ dz/dt = xy - 4z \end{cases} \tag{7.1}$$

(2) driven Duffing oscillator described by

$$\begin{cases} dx/dt = y \\ dy/dt = -0.05y + 0.5x - 0.5x^3 + 7.5\cos(z) \\ dz/dt = 1 \end{cases} \tag{7.2}$$

(3) chaotic Rössler oscillator defined by

$$\begin{cases} dx/dt = -y - z \\ dy/dt = x + 0.2y \\ dz/dt = 0.2 + z(x - 5.7) \end{cases} \tag{7.3}$$

To visualize the PSCNs, we use the Kamada-Kawai spring embedding algorithm [13]. The results are shown in Fig. 7.1. For each example, we observe some resemblance between the PSCN and the original chaotic attractor. After PSCNs are obtained, we employ the software FANMOD to detect network motifs. The software is developed using the Wernicke algorithm proposed in Refs. [14, 15]. To be concrete, we focus on six different motifs of size four (i.e., four nodes) and calculate their frequencies of occurrence, as shown in Fig. 7.2. We observe a common feature among three examples: the motif distribution is highly heterogeneous. For example, the frequencies of motifs A and B are apparently much higher than those of others (e.g., motifs E and F). The heterogeneity originates from the UPOs embedded in the chaotic attractor. It has been known that, while the infinite set of periodic orbits embedded in a chaotic set are all unstable, their stabilities as determined by the corresponding largest eigenvalues are typically quite heterogeneous. In fact, a small set of UPOs can be significantly less unstable than the others. In the phase space, a UPO appears as a closed loop. A chaotic trajectory tends to spend substantially more time near weakly unstable orbits. As a result, there can be many recurrences near such a UPO, giving rise to a cluster in the corresponding PSCN, as can be seen from Figs. 7.1 and 7.2. We have also tested numerically that the motif distributions are robust with respect to variation in the network size, insofar as there are at least a few thousands of nodes.

Will weak noise affect the heterogeneous nature in the motif distribution of PSCN from chaotic time series? To address this question, we add Gaussian white

Fig. 7.1 PSCN of 2000
nodes from (**a**) Lorenz system
t = 20; **b** Duffing system
with t = 100; and **c** Rössler
system with t = 200. The
resulting networks are drawn
by the software UCINET and
PAJEK [16, 17]

(a)

(b)

(c)

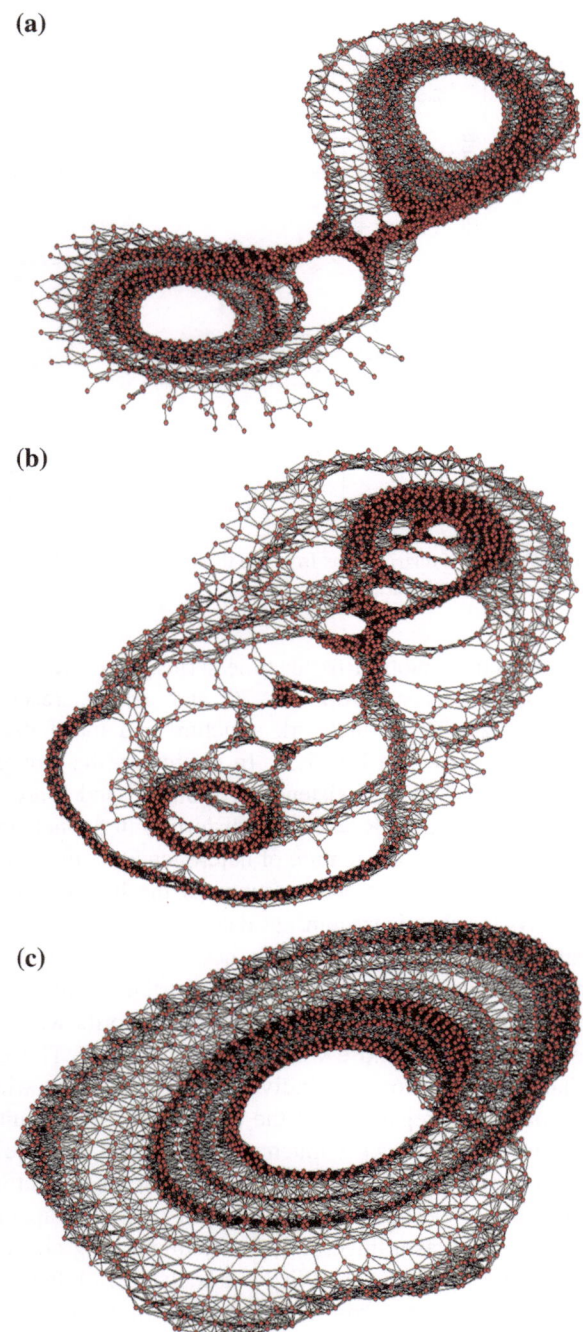

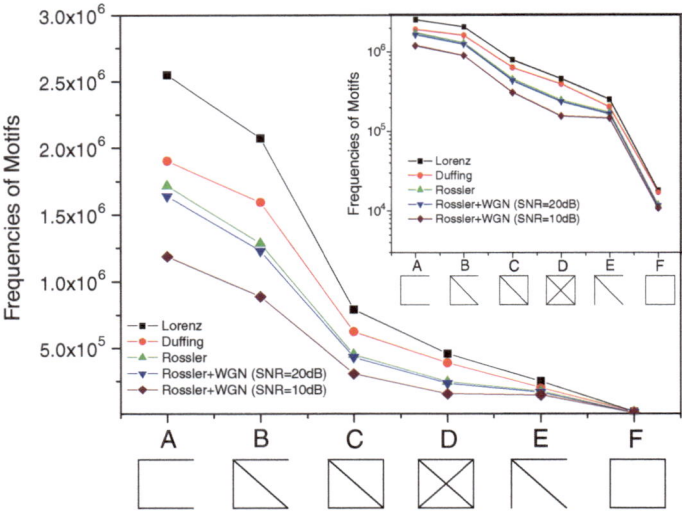

Fig. 7.2 Motif distributions from the three chaotic PSCNs in Fig. 7.1. The *inset* shows the motif distributions on logarithmic scale

noise to the chaotic Rössler time series to generate two time series whose signal-to-noise ratios (SNRs) are 20 and 10 dB, respectively. We find that, for SNR = 20 dB, the network structure and motif distribution are essentially the same, as shown in Fig. 7.3a. In particular, the strong heterogeneity in the motif distribution and the existence of motifs E and F are unchanged. For larger noise amplitude, i.e., SNR = 10 dB, distortion in the network structure arises, as shown in Fig. 7.3b, but the feature of heterogeneity still remains. It is noteworthy that the presence of noise tends to suppress the heterogeneity of motif distributions in PSCNs for chaotic systems, as demonstrated in Fig. 7.3b. In particular, under noise an unstable periodic orbit may not close on itself and the transitions among different orbits are randomized. The distribution of all motifs will be influenced by the randomization but, statistically, the motifs with higher frequencies are influenced more than those with lower frequencies. This can be understood by noting that the distortions to individual motifs occur with approximately the same probabilities, regardless of the motif type. As a result, the motif with the highest frequency (motif A) is much more sensitive to noise than other motifs, reducing considerably its absolute number. Despite this effect under strong noise, for weak noise, the heterogeneous distribution of motifs appears to be robust, which can then be used as a fingerprint for distinguishing chaotic time series.

Figure 7.4 shows the PSCNs associated with the three distinct types of inclined oil-water flow patterns. The respective motif distributions are shown in Fig. 7.5. We observe a common feature among the distributions: heterogeneity. The similarity to the motif distributions from typical chaotic systems leads us to speculate that the dynamics underlying the three flow patterns may be chaotic. While the

(a)

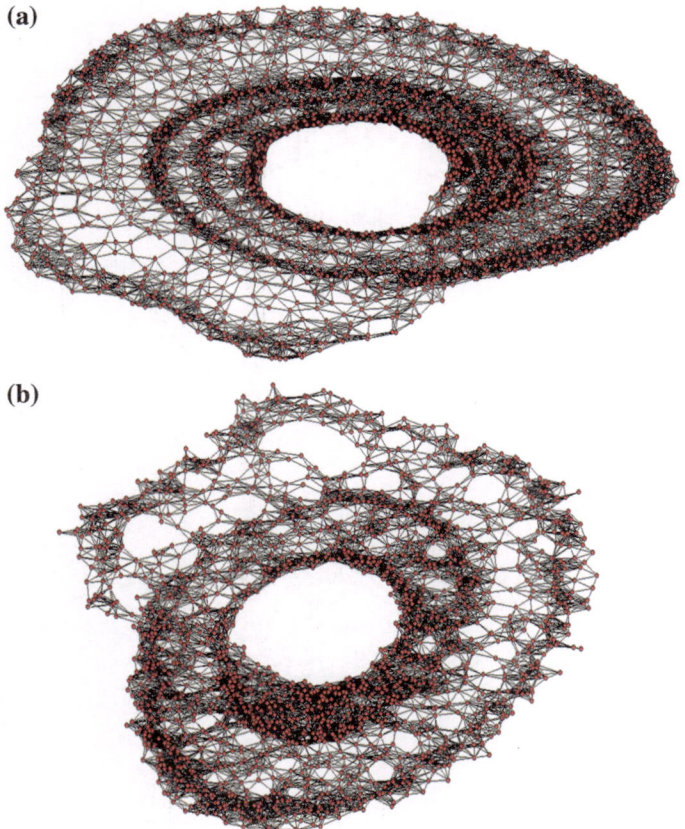

(b)

Fig. 7.3 PSCN of 2000 nodes from Rössler chaotic time series corrupted by Gaussian white noise. **a** SNR = 20 dB and **b** SNR = 10 dB

observed motif distributions all appear heterogeneous, the degrees of heterogeneity are apparently different for different flow patterns, where the transitional flow exhibits the most heterogeneous distribution. To give credence to our proposition that the dynamics of inclined oil-water two-phase flows are chaotic, we have computed the maximal Lyapunov exponent (MLE) from time series by using a standard method [18]. The MLE is computed by abandoning the first 1000 transient data points and using the following 10,000 data points. The estimate MLEs are 0.084 ± 0.008, 0.055 ± 0.003, and 0.037 ± 0.003 for TF, CT, and PS flows, respectively. All MLEs are positive, suggesting strongly the chaotic nature of the underlying flows. More remarkably, the flow with the largest value of MLE corresponds to the most heterogeneous PSCN motif distribution.

In PS flow pattern, due to the approximately periodic switch between oil-plug and water-plug movements, the flow also exhibits approximately regular dynamics to some extent, leading to the smallest MLE value and, consequently, to the

(a)

(b)

(c)

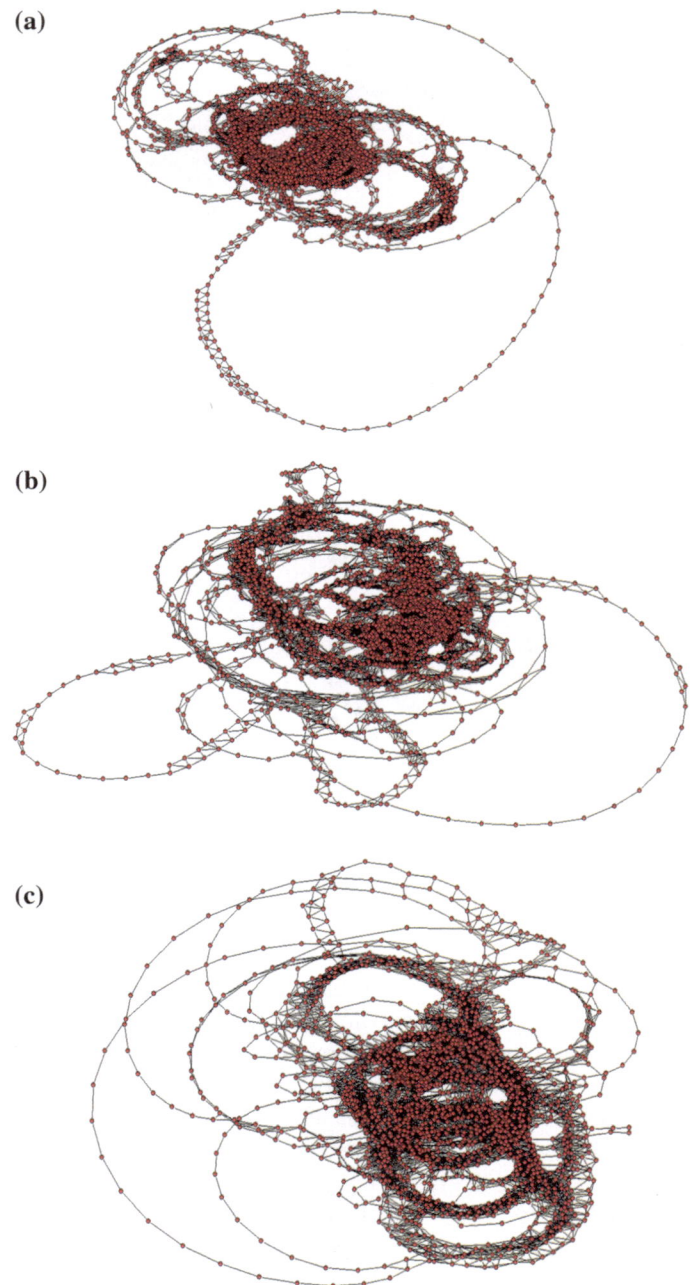

Fig. 7.4 PSCNs associated with three distinct types of two-phase flow patterns: **a** PS flow ($U_{so} = 0.01899$ m/s, $U_{sw} = 0.00615$ m/s), **b** CT flow ($U_{so} = 0.09813$ m/s, $U_{sw} = 0.00605$ m/s), and **c** transitional flow ($U_{so} = 0.12569$ m/s, $U_{sw} = 0.00614$ m/s). Each network is constructed by setting threshold r_c to be 20 % of the rms value of the measured conductance fluctuating signal and contains 2000 nodes

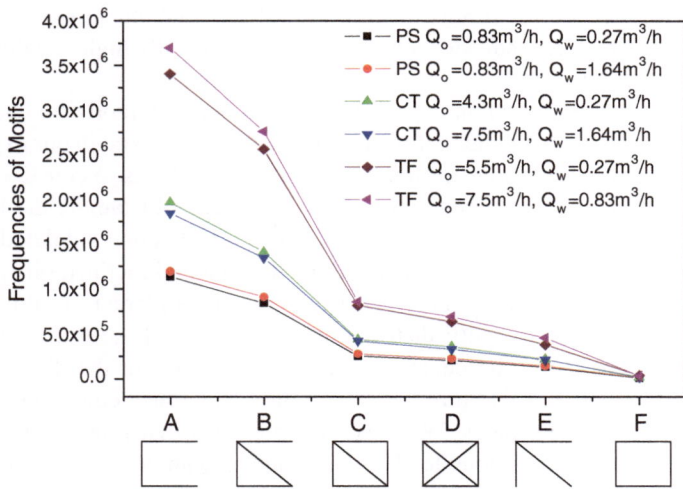

Fig. 7.5 PSCN motif distributions of three distinct flow patterns (two signals for each flow pattern). The *inset* shows the motif distributions in logarithmic scale

weakest heterogeneity in the motif distribution as characterized by, e.g., low frequencies of occurrences of motifs A and B as compared with those of the CT and transitional flows. In CT flow pattern, oil disperses in the continuous water in the form of discrete well rounded droplets of mostly small to medium size. Because of the density difference between oil and water, the droplets pass the upper regions of the pipe in an uninterrupted sequence with nearly uniform vertical spread. The prominent characteristic of CT flow patterns is the local countercurrent flow of water, which usually occurs at the bottom side of the pipe. The countercurrent phenomenon is due to the increasing magnitude of the gravitational component in the direction opposite to the main flow, which partially overcomes the linear momentum of the water phase. Reflected in the motif distribution, we observed that the frequencies of motifs A and B increase as the flow pattern switches from PS to CT. Indeed, we find that the values of MLE for CT flows tend to be larger than those for PS flows. Transitional flows arise at moderate oil superficial velocities associated with low to moderate superficial water velocities, which appear in the regime between the water-dominated and oil-dominated flow patterns. In the TF regime, oil regions form at the top side of the pipe, while water exists at the bottom of the pipe with a few recirculating oil droplets. In the middle region the oil and water phases appear alternatively as a continual phase. TF patterns thus appear more complex than PS and CT flows. Indeed, we find that the values of the MLE associated with transitional flows are generally larger. As can be seen from Fig. 7.5, the motif distributions associated with the TFs are more heterogeneous than those with PS and CT flows. Moreover, from Figs. 5.11 and 7.5, we can see that TF flows possess the highest degree of complexity while PS flows have the lowest value, with values from the CT flows lying in between, providing support for our results based on motif distributions in that more heterogeneous distributions indicate behaviors associated with higher values of the complexity measures.

Thus, we have proposed an approach based on network motifs to characterize three water-dominated inclined oil-water flow patterns observed in our experiment. Given a measured time series from a flow pattern, our idea is to construct a phase space and then a network based on a distance metric, and examine the frequency distribution of a number of network motifs. The method is first validated by using classical chaotic systems and then applied to experimental inclined oil-water two-phase flows. We find that the three water-dominated countercurrent flow patterns all exhibit heterogeneous motif distributions, mimicking those from chaotic systems. More remarkably, the degrees of the heterogeneity in the distribution are distinctly different for the three types of flow patterns, consistent with the Lyapunov-exponent estimates. We have also observed that flows with more heterogeneous motif distributions tend to have higher complexity measures. These results suggest the power of network motifs to characterize and distinguish complex flow patterns.

References

1. Z.K. Gao, N.D. Jin, W.X. Wang, Y.C. Lai, Motif distributions in phase-space networks for characterizing experimental two-phase flow patterns with chaotic features. Phys. Rev. E **82**(2), 016210 (2010)
2. R. Milo, S. Shen-Orr, S. Itzkovitz, N. Kashtan, D. Chklovskii, U. Alon, Network motifs: simple building blocks of complex networks. Science **298**(5594), 824–827 (2002)
3. R. Milo, S. Itzkovitz, N. Kashtan, R. Levitt, S. Shen-Orr, I. Ayzenshtat, M. Sheffer, U. Alon, Superfamilies of evolved and designed networks. Science **303**(5663), 1538–1542 (2004)
4. E. Yeger-Lotem, S. Sattath, N. Kashtan, S. Itzkovitz, R. Milo, R.Y. Pinter, U. Alon, H. Margalit, Network motifs in integrated cellular networks of transcription-regulation and protein–protein interaction. Proc. Natl. Acad. Sci. USA **101**(16), 5934–5939 (2004)
5. O. Rahat, U. Alon, Y. Levy, G. Schreiber, Understanding hydrogen-bond patterns in proteins using network motifs. Bioinformatics **25**(22), 2921–2928 (2009)
6. A. Sharma, S. Chavali, R. Tabassum, N. Tandon, D. Bharadwaj, Gene prioritization in type 2 diabetes using domain interactions and network analysis. BMC Genomics **11**, 84 (2010)
7. S. Mangan, U. Alon, Structure and function of the feed-forward loop network motif. Proc. Natl. Acad. Sci. USA **100**(21), 11980–11985 (2003)
8. N. Kashtan, S. Itzkovitz, R. Milo, U. Alon, Topological generalizations of network motifs. Phys. Rev. E **70**(3), 031909 (2004)
9. K. Baskerville, P. Grassberger, M. Paczuski, Graph animals, subgraph sampling, and motif search in large networks. Phys. Rev. E **76**, 036107 (2007)
10. C.G. Li, Functions of neuronal network motifs. Phys. Rev. E **78**, 037101 (2008)
11. G. Chechik, E. Oh, O. Rando, J. Weissman, A. Regev, D. Koller, Activity motifs reveal principles of timing in transcriptional control of the yeast metabolic network. Nat. Biotechnol. **26**(11), 1251–1259 (2008)
12. X. Xu, J. Zhang, M. Small, Superfamily phenomena and motifs of networks induced from time series. Proc. Natl. Acad. Sci. USA **105**, 19601–19605 (2008)

13. T. Kamada, S. Kawai, An algorithm for drawing general undirected graphs. Inform. Process. Lett. **31**(1), 7–15 (1989)
14. S. Wernicke, F. Rasche, FANMOD: a tool for fast network motif detection. Bioinformatics **22**(9), 1152–1153 (2006)
15. S. Wernicke, Efficient detection of network motifs. IEEE-ACM Trans. Comput. Biol. Bioinform. **3**(4), 347–359 (2006)
16. S.P. Borgatti, M.G. Everett, L.C. Freeman, *UCINET for Windows: Software for Social Network Analysis* (Analytic Technologies, Harvard, 2002)
17. W. Nooy, A. Mrvar, V. Batagelj, *Exploratory Social Network Analysis with PAJEK* (Cambridge University Press, New York, 2005)
18. A. Wolf, J.B. Swift, H.L. Swinney, J.A. Vastano, Determining Lyapunov exponents from a time series. Physica D **16**(3), 285–317 (1985)

Chapter 8
Directed Weighted Complex Network for Characterizing Gas-Liquid Slug Flow

8.1 Methodology

Recently, we have proposed a framework for inferring a directed weighted complex network from a time series [1]. We here introduce the analytical framework as follows: we start from construction of the *Directed weighted complex network* (DWCN). Our first step is phase space reconstruction. Given a time series $z(it)$ ($i = 1, 2..., M$), where t is the sampling interval and M is the sample size, we construct a sequence of phase-space vectors according to the standard delay-coordinate embedding method [2–4]:

$$\vec{X}_k = \{x_k(1), x_k(2), \ldots, x_k(m)\}$$
$$= \{z(kt), z(kt + \tau), \cdots, z(kt + (m - 1)\tau)\} \quad (5.1)$$

where τ is the delay time, m is the embedding dimension, $k = 1, 2,..., N$, and $N = M - (m - 1)\tau/t$ is the total number of vector points in the reconstructed phase space. To construct a network, we then regard each vector point as a node and use the phase-space distance to determine the edges. Given two vector points $\vec{X}_i$ and $\vec{X}_j (i > j)$, the phase-space distance is defined to be

$$d_{ij} = \sum_{n=1}^{m} \left\| X_i(n) - X_j(n) \right\| \quad (5.2)$$

where $X_i(n) = z(i + (n - 1)\tau)$ is the nth element of $\vec{X}_i$. This generates, for all nodes (vector points) in the network, a distance matrix $\mathbf{D} = (d_{ij})$ where $i > j$. By choosing a critical threshold value r_c, we obtain the connections of the network: an edge connecting node i and j ($i > j$) exists if $|d_{ij}| \leq r_c$; while there is no edge between i and j if $|d_{ij}| > r_c$. We regard the time $(i - j) * t$ as the weight of an edge that connected nodes i and j and the edge direction is from node i to j. Finally, we obtain the weight matrix $\mathbf{W} = (w_{ij})$, where $w_{ij} = 0$ means node i and j are not connected, otherwise, $w_{ij} \neq 0$ implies an edge from node i to j exists and the edge

Z.-K. Gao et al., *Nonlinear Analysis of Gas-Water/Oil-Water Two-Phase Flow in Complex Networks*, SpringerBriefs on Multiphase Flow, DOI: 10.1007/978-3-642-38373-1_8, © The Author(s) 2014

weight is $w_{ij} = (i - j) * t$. The topology of the reconstructed DWCN is determined entirely by $\mathbf{W}$.

A key issue in extracting DWCN from time series is then the choice of the critical threshold r_c. In this chapter, we exploit normalized maximum size of subgraph to determine the critical threshold. Take the Tent map

$$f(x) = \begin{cases} 2x, & \text{if } x < 1/2 \\ 2(1 - x), & \text{if } x \geq 1/2 \end{cases} \tag{5.3}$$

as an example, we here theoretically demonstrate how to properly select the critical threshold. The locations of the periodic orbits of the tent map can be obtained explicitly. At each iteration, the map has two line segments of slope 2 and -2 in the unit square of the plane x_{n+1} versus x_n. The pth-iterated map has 2^P line segments in the unit square. Fixed points of the pth-iterated map, which contain all periodic orbits of period p, are located at cross points of these 2^P line segments with the line $x_{n+1} = x_n$. Thus we have the following set of points that belong to different periodic orbits of period p:

$$X_P(j) = \begin{cases} 2j/(2^P + 1), & \text{if } j = 1, 2, 3 \cdots, 2^{P-1} \\ 2j/(2^P - 1), & \text{if } j = 1, 2, 3, \cdots, 2^{P-1} - 1 \end{cases} \tag{5.4}$$

Starting with one such point, one can obtain the remaining p-1 points on the orbit by iterating the tent map. We obtain all the orbit points from period 1 to 11 considering the fact that the chaotic dynamics mainly focus on the unstable periodic orbits of low periods. Then we calculate the minimum distance from the point i on the orbit q_l of period P_l to the points of other orbits

$$L_{P_l}^{q_l}(i) = \min \left(\left\{ \left[X_{P_l}^{q_l}(i) - X_{P_k}^{q_k}(j) \right]^2 + \left[f\left(X_{P_l}^{q_l}(i) \right) - f\left(X_{P_k}^{q_k}(j) \right) \right]^2 \right\}^{1/2} \middle| \begin{array}{l} P_k = 1, 2, \cdots, 11 \\ q_k = 1, 2, \cdots, 2^{P_k} - 1 \\ (P_k, q_k) \neq (P_k, q_l) \end{array} \right) \tag{5.5}$$

where $2^{p_k-1} - 1$ is the number of orbit of period P_k. $L_{P_l}^{q_l}(i)$ is the minimum distance for point i. Then we calculate the minimum distance for all points on the orbits of different periods and select the maximum value of the obtained minimum distances

$$r_C = \max \left(L_{P_l}^{q_l}(i) \middle| P_l = 1, 2, \cdots, 11 \quad q_l = 1, 2, \cdots, 2^{P_l} - 1 \quad i = 1, 2, \cdots, P_l \right)$$
$$= 0.00155 \tag{5.6}$$

as network critical threshold in that it is the smallest one that can guarantee the skeleton of the network. So, we have theoretically calculated the critical threshold from the equations for Tent map system.

To cast light into the time series of length 8,000 from tent map, we now carry out the *normalized maximum size of subgraph* (NMSS) distribution to determine the critical threshold, as follows. We reconstruct the network and study the distributions of NMSS with respect to the variation in threshold. For a network from time series, the increase of NMSS will reach the maximum rate when the threshold approaches the smallest value that can preserve the network skeleton. By analysis, we find that when r_c approaches 0.00175 the NMSS reaches maximum increase rate. Thus, the $r_c = 0.00175$ is the critical threshold for tent map and the result is in agreement with that obtained from theoretical calculation $r_c = 0.00155$. Furthermore, we use the NMSS distribution study the 2x mod 1 map and the results indicate that, consistently with the one theoretically calculated from equations $r_c = 0.0021$, the critical threshold obtained from NMSS distribution is $r_c = 0.00205$, confirming that the NMSS distribution could be a faithful approach for determining the network critical threshold.

To better demonstrate our DWCN method, without loss of generality, we focus on time series (8,000 points) from four typical dynamic systems: (1) chaotic Lorenz system [5] given by $\dot{x} = 16(y - x)$, $\dot{y} = x(45.92 - z) - y$, and $\dot{z} = xy - 4z$, (2) chaotic Rössler oscillator [6] defined by $\dot{x} = -y - z$, $\dot{y} = x + 0.2y$, and $\dot{z} = 0.2 + z(x - 5.7)$, (3) chaotic Rössler oscillator + Gauss white noise, (4) Gauss white noise. We first investigate the NMSS distributions for the four different systems, as shown in Fig. 5.1, where the *rms* is the root mean-square value of the time series, C is the threshold coefficient and C_c is critical threshold coefficient and $r_c = C_c * rms$. Then we proceed to construct DWCN and investigate the degree and weight correlation distribution of the result networks. We find that the distributions of the DWCN from different systems all can be well fitted with a power law $W = D^\gamma$, where W is the input/output weight, D is the input/output degree and γ is the correlation exponent. In fact, the power law distribution originates from the nonlinear accumulation of node weight, i.e., the node weight not linearly but nonlinearly increases with the node degree. Figure 5.2 shows the distributions of degree and weight correlation exponent γ with respect to the variation in threshold coefficient. As can be seen from Figs. 5.1 and 5.2, time series with different dynamics exhibit distinct C_c value and γ distribution. We now consider how these distinctions are related to the configurations of the *unstable periodic orbits* (UPOs) embedded in a chaotic attractor. For a chaotic system, e.g., Lorenz and Rössler system, its trajectory tends to approach different UPOs in different time, which constitute the "skeleton" of the attractor. It has been known that, while the infinite set of periodic orbits embedded in a chaotic set are all unstable, their stabilities as determined by the corresponding largest eigenvalues are typically heterogeneous. In fact, a small set of UPOs can be significantly less unstable than others. In the phase space, a UPO appears as a closed loop. A chaotic trajectory tends to spend substantially more time near weakly unstable orbits. As a result, there can be many recurrences near such a UPO, giving rise to a dense orbit region. Since a chaotic attractor contains infinitely UPOs, there will be multiple dense orbit regions in the phase space, resulting in small critical threshold coefficient C_c, as can be seen from Fig. 5.1. Moreover, for the DWCN, with the increase of C, dense orbit regions will give rise to high degree nodes and quick

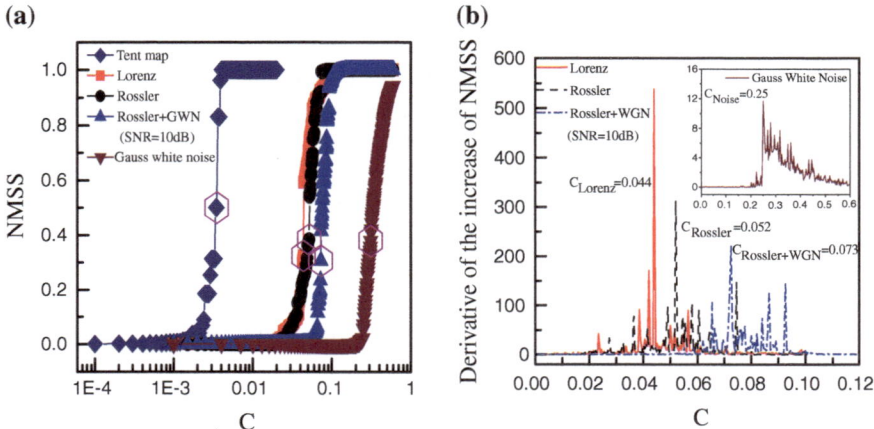

Fig. 5.1 a Distributions of NMSS with respect to the variation in threshold coefficient, where C is the threshold coefficient, *rms* is the root mean-square value of the time series and $r_c = C_c * rms$. **b** Derivative of the increase of NMSS. The critical C_c is marked by magenta hexagon where $C_{Tent\ map} = 0.0035$, $C_{Lorenz} = 0.044$, $C_{Rössler} = 0.052$, $C_{Rössler+GWN} = 0.073$, $C_{GWN} = 0.25$

accumulation of node weight, i.e., large degree and weight correlation exponent γ. Because of the existence of multiple dense orbit regions induced from UPOs of different stability, the power law exponent will monotonically increase with C for a large range, as shown in Fig. 5.2. Thus, the distributions of γ exponent represent the hierarchical structure of the UPOs. It quantifies the richness of the UPOs of the chaotic attractor and serves to characterize the chaotic dynamics in terms of the stability of the UPOs. In contrast, for the Gauss white noise, since it possesses strong stochastic property and there does not exist any dense orbit regions associated with UPOs, a large critical threshold coefficient C_c (nearly one order of magnitude larger than that of chaotic system) is required to form a noisy DWCN and its exponent γ that is smaller than that of chaotic system is independent of C, as shown in Figs. 5.1 and 5.2.

Will noise affect the properties of DWCN induced from UPOs? To address this question, we add Gauss white noise to the chaotic Rössler time series to generate a time series whose SNR (signal-to-noise ratios) is 10 dB. We find that, despite the intense influence of noise, the critical C_c and the correlation γ exponent distribution are essentially the same, as shown in Figs. 5.1 and 5.2, indicating the robustness of the DWCN properties. Therefore, the critical threshold coefficient C_c and the γ exponent distribution can then be used as a fingerprint for distinguishing different dynamical regimes associated with UPOs.

Detecting unstable periodic orbits from time series is a fundamental problem of continuing interest in a wide variety of fields [7–9]. A question is, can unstable periodic orbits be detected from chaotic time series in terms of DWCN? We shall demonstrate that the answer to the above question is affirmative. Actually, for the chaotic DWCN, its critical threshold $r_c = C_c * rms$ obtained from NMSS is small

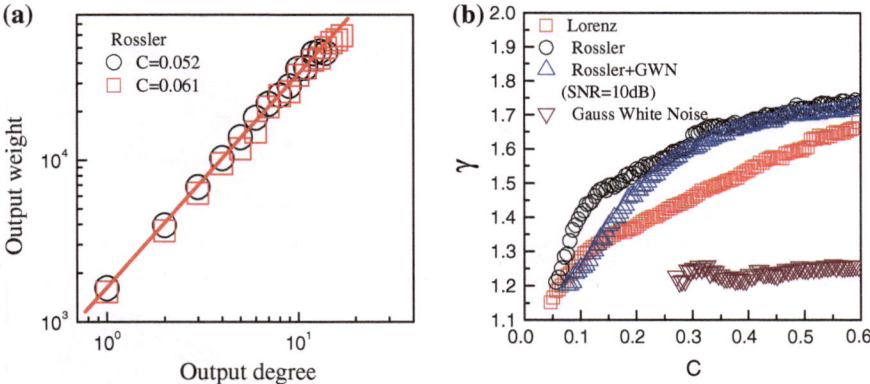

Fig. 5.2 a The power law distribution for Rössler system where the data points are obtained by averaging over all weights with the same degree. **b** The distributions of degree and weight correlation exponents with respect to the variation in threshold coefficient. Note that the input and output distribution are intrinsic the same, so we here only show the results of output degree and weight correlation distributions

enough to guarantee the connected two nodes i and $j(i > j)$ are recurrent and the corresponding recurrence time is just the edge weight $w_{ij} = (i - j) * t$. According to the idea of LK method [7], a recurrent node is not necessarily a component of a periodic orbit of period $w_{ij} = (i - j) * t$, but if a particular recurrence time w_{ij} appears frequently in the constructed DWCN, it is likely that the corresponding recurrent nodes are close to periodic orbits of period w_{ij}. The idea is then to construct a histogram of the edge weights, i.e., recurrence times, and identify peaks in the histogram. Nodes that occur frequently with are taken to be, approximately, components of the periodic orbits. By analysis, we find that the UPOs of low periods can be successfully detected from chaotic time series via DWCN, as shown in Fig. 5.3a. In addition, we examine the resistance of the method by inserting noise into time series ($SNR = 20$ dB), and find that although distortion in the orbits arises, UPOs still can be detected in terms of DWCN (see Fig. 5.3b), indicating our method is robust against noise. These results demonstrate that the DWCN approach could also be an effective tool for detecting unstable periodic orbits of different periods from chaotic time series.

8.2 Characterizing Chaotic Dynamic Behavior in Slug Flow

We have used the proposed directed weighted complex network to characterize gas-liquid slug flow [10]. This DWCN view of time series reveals interesting results and yields measures that help our understanding of the chaotic dynamics.

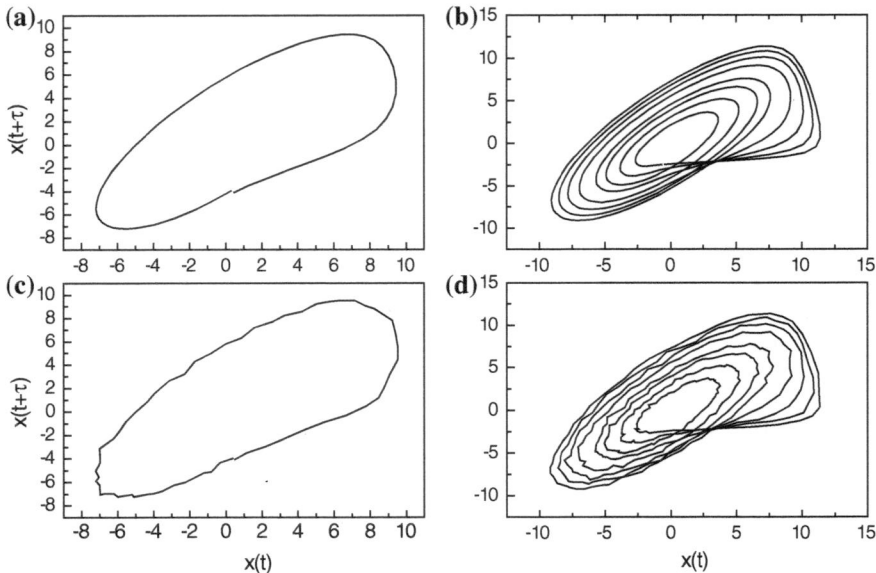

Fig. 5.3 For Rössler system: **a, b** a period-1 and period-8 orbits; For a noisy Rössler system ($SNR = 20$ dB): **c, d** a period-1 and period-8 orbits

Now we will demonstrate the effectiveness of our method in characterizing the complex dynamics underlying gas-liquid slug flow. We first calculate the NMSS distribution of the three gas-liquid flow patterns from 30 experimental measurements (Fig. 5.4 displays the NMSS distributions for bubble flow, slug flow and churn flow) and then demonstrate in Fig. 5.5 that critical threshold coefficient C_c for three typical types of flow patterns are located in three distinct regions without any overlap. Note that there exists an abrupt transition of critical threshold coefficient between bubble and slug flow, indicating that the dynamical features of the two flow patterns are quite different, consistent with existent knowledge about the phase flows [11]. These results demonstrate that the DWCN based approach is capable of distinguishing the three typical flow patterns.

Figure 5.5 indicates that the critical threshold coefficient C_c of slug flow is 0.062 ± 0.003, i.e., of the same order of magnitude as that of chaotic Rössler system, while the C_c of bubble flow is 0.233 ± 0.012, i.e., of the same order of magnitude as that of Gauss white noise system, and the C_c of churn flow which is 0.181 ± 0.008 lies between them. This leads us to speculate that the phase space of signals measured from slug flow may contain numbers of UPOs, while other two flow patterns do not. To give credence to our speculation, we proceed to construct DWCN from two-phase flow measurement signals, as shown in Fig. 5.6, and investigate the degree-strength correlation distribution of the DWCN corresponding to different flow patterns, as shown in Fig. 5.7. As can be seen, the degree-strength correlation of different flow patterns all can be characterized by a

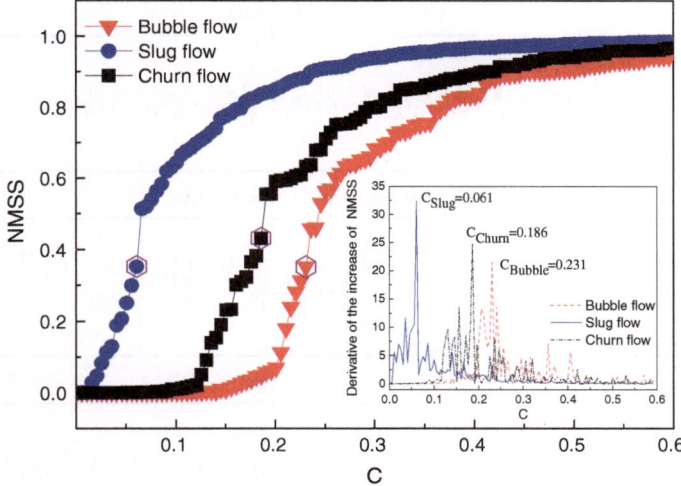

Fig. 5.4 Distributions of NMSS with respect to the variation in threshold coefficient for three different flow patterns. The inset indicates the derivative of the increase of NMSS. The critical threshold coefficient C_c is marked by magenta hexagon where $C_{Bubble} = 0.231$, $C_{Slug} = 0.061$, $C_{Churn} = 0.186$

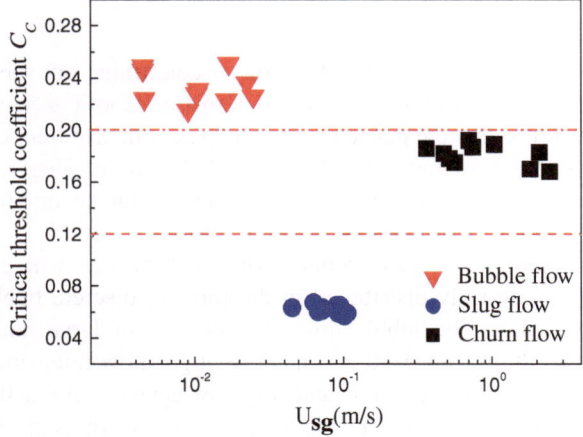

Fig. 5.5 Distributions of critical threshold coefficient C_c on semi-log scale for 30 measurement signals belonging to three patterns with increase of the gas superficial velocity

power law exponent. Figure 5.8 shows the distributions of degree-strength correlation exponent γ with respect to the variation in threshold coefficient. As can be seen, because of the existence of multiple dense orbit regions induced from UPOs of different stability, the power law exponent of slug flow monotonously increases with C for a large range, while bubble and churn flow are almost independent of

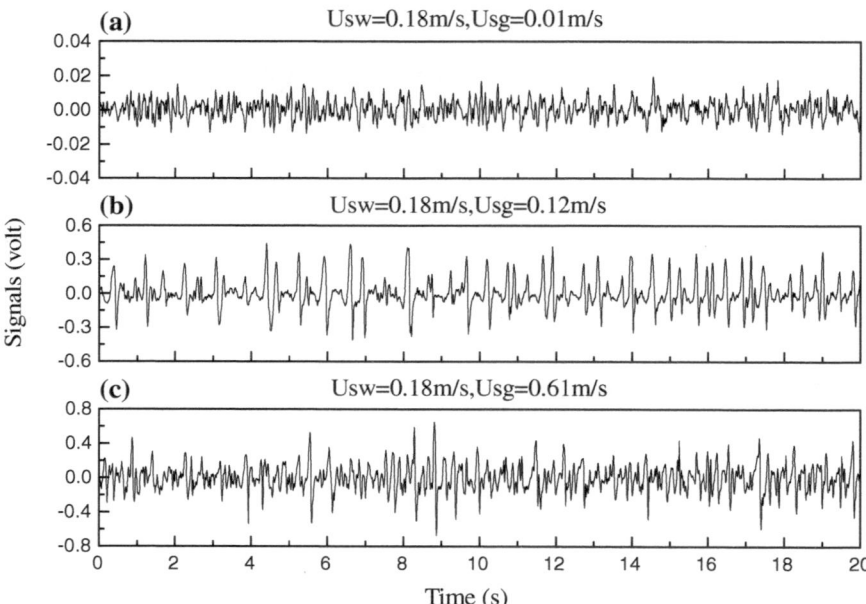

Fig. 5.6 The conductance fluctuating signals in three flow patterns. **a** Bubble flow; **b** Slug flow; **c** Churn flow

C due to the non-existence of the UPOs, strongly confirming our speculation. Daw et al. [12–14] investigated two-phase flow in terms of chaotic time-series analysis methods and believed that the existence of an apparent intermittent transition from periodic to possibly low-dimensional chaotic behavior in slugging regimes in fluidized beds. Our method quantitatively characterizes the chaotic features of gas-liquid slug flow in terms of unstable periodic orbits.

In fact, gas-liquid bubble flow occurs at low gas flow rates where the gas phase is approximately uniformly distributed in the form of discrete bubbles in a continuum of liquid phase. In bubble flow, the motions of large number of small bubbles are rather stochastic. With an increase in gas flow rate, small air bubbles begin to coalesce to form a large one and slug flow appears. In this flow, gas phase exists in the form of large bullet shaped bubble, also known as gas slug or Taylor bubble, whose diameters almost equal the pipe diameter. The liquid slug area between two Taylor bubbles is filled with small bubbles. The quasi-periodic oscillation behavior between slugs can be characterized by unstable periodic orbits in the reconstructed phase space. This has led to the fact that the power law exponent γ of slug flow will monotonously increase with C for a large range, while the γ exponent of bubble flow is independent of C, indicating no UPOs exist in the DWCN from bubble flow. Actually, the orbits of slug flow are basically composed of one big loop and one small loop, as can be seen in Fig. 5.9. The orbit develops from small loop to big loop represents intermittent quasi-periodic oscillation

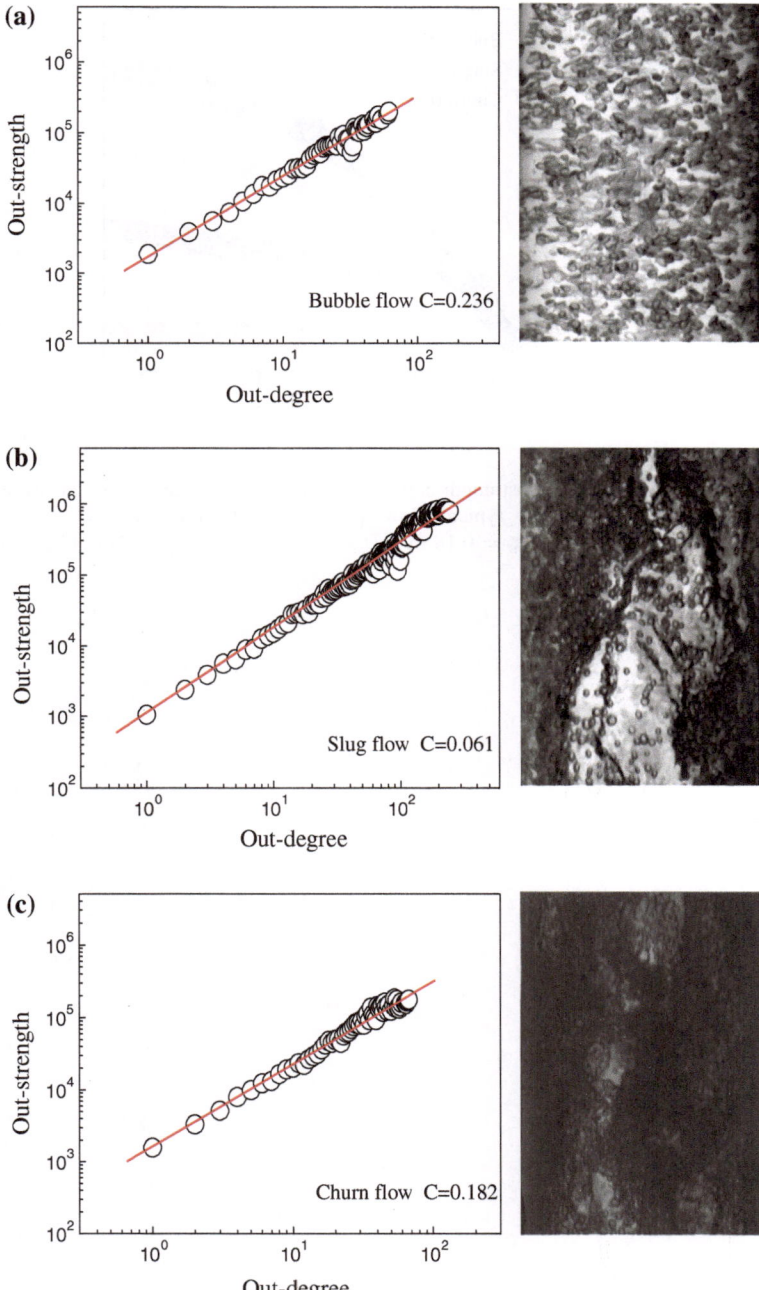

Fig. 5.7 The power law distribution for three typical gas-liquid two-phase flow patterns where the data points are obtained by averaging over all strength with the same degree. **a** *Bubble flow* ($U_{sw} = 0.18$ m/s, $U_{sg} = 0.01$ m/s); **b** *Slug flow* ($U_{sw} = 0.18$ m/s, $U_{sg} = 0.12$ m/s); **c** *Churn flow* ($U_{sw} = 0.18$ m/s, $U_{sg} = 0.61$ m/s). The *right panel* indicates the dynamic snapshots for three typical gas–liquid flow patterns recorded by High-speed VCR

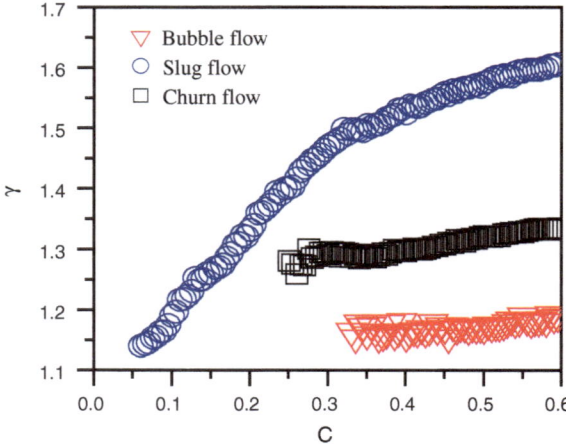

Fig. 5.8 Distributions of degree-strength correlation exponents with respect to the variation in threshold coefficient for three typical flow patterns. **a** *Bubble flow* ($U_{sw} = 0.18$ m/s, $U_{sg} = 0.01$ m/s); **b** *Slug flow* ($U_{sw} = 0.18$ m/s, $U_{sg} = 0.12$ m/s); **c** *Churn flow* ($U_{sw} = 0.18$ m/s, $U_{sg} = 0.61$ m/s)

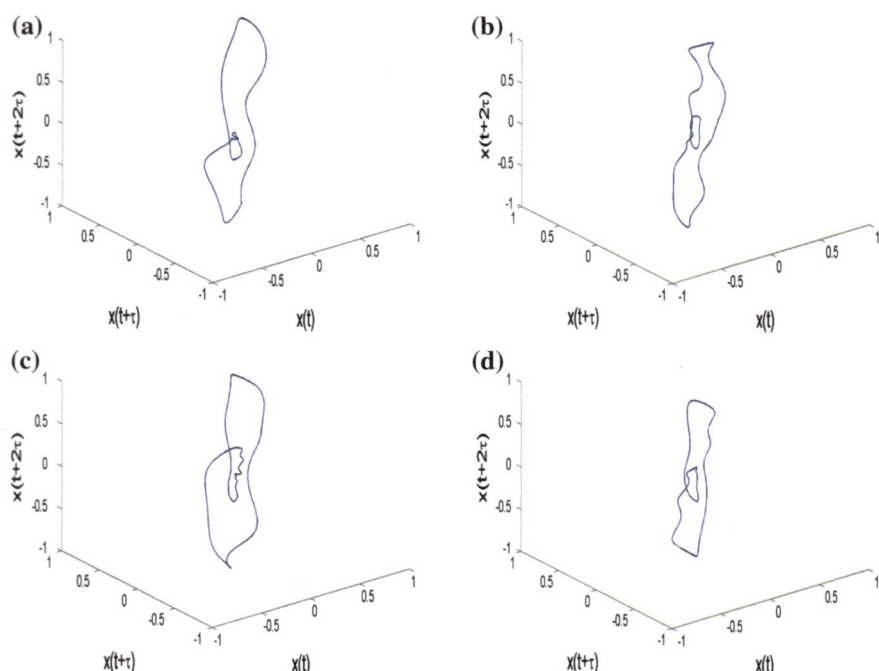

Fig. 5.9 The UPO from slug flow under different flow conditions detected by DWCN. **a** $U_{sw} = 0.0226$ m/s, $U_{sg} = 0.0679$ m/s; **b** $U_{sw} = 0.0226$ m/s, $U_{sg} = 0.1019$ m/s; **c** $U_{sw} = 0.0453$ m/s, $U_{sg} = 0.0951$ m/s; **d** $U_{sw} = 0.18$ m/s, $U_{sg} = 0.12$ m/s

behavior between liquid slug containing small gas bubbles and large gas slug. Furthermore, we find that for the UPOs from slug flows corresponding to different flow conditions, although the "skeleton" is the same, the details varies with flow parameter, as shown in Fig. 5.9a–d, indicating that the detailed changes in slug flow oscillations under different flow conditions also could be uncovered in terms of UPOs. Churn flow that happens at high gas flow rates is composed of discrete gas phase and continuous liquid phase of high turbulent kinetic energy. Churn flow is a highly disordered flow, due to the influence of turbulence effect, its oscillation behavior is not so obvious as slug flow, resulting in the fact that the critical threshold coefficient C_c of churn flow is much smaller than bubble flow but larger than slug flow. Comparing with the DWCN from slug flow, the γ exponent of churn flow is almost independent of C, indicating no UPOs can be detected from the DWCN of churn flow. Therefore, from measured signals to bubble dynamics, we associate the macroscopic behavior with the inherently complex microscopic dynamics in terms of directed weighted complex network.

References

1. Z.K. Gao, N.D. Jin, A directed weighted complex network for characterizing chaotic dynamics from time series. Nonlinear Anal. Real World Appl. **13**, 947–952 (2012)
2. N.H. Packard, J.P. Crutchfield, J.D. Farmer, Geometry from a time series. Phys. Rev. Lett. **45**(9), 712–716 (1980)
3. F. Takens, *Dynamical Systems and Turbulence, Lecture Notes in Mathematics*, vol. 898 (Springer, New York, 1981), pp. 366–381
4. T. Sauer, J.A. Yorke , M. Casdagli, Embedology. J. Stat. Phys. **65**, 579–616 (1991)
5. E.N. Lorenz, Deterministic nonperiodic flow. J. Atmos. Sci. **20**, 130–148 (1963)
6. O.E. Rössler, An equation for continuous chaos. Phys. Lett. A **57**, 397–398 (1976)
7. D.P. Lathrop, E.J. Kostelich, Characterization of an experimental strange attractor by periodic orbits. Phys. Rev. A **40**, 4028–4031 (1989)
8. P. So, E. Ott, S.J. Schiff, D.T. Kaplan, T. Sauer, C. Grebogi, Detecting unstable periodic orbits in chaotic experimental data. Phys. Rev. Lett. **76**, 4078–4705 (1996)
9. M. Dhamala, Y.C. Lai, E.J. Kostelich, Detecting unstable periodic orbits from transient chaotic time series. Phys. Rev. E **61**, 6485–6489 (2000)
10. Z.K. Gao, N.D. Jin, Characterization of chaotic dynamic behavior in the gas-liquid slug flow using directed weighted complex network analysis. Physica A **391**, 3005–3016 (2012)
11. Z.K. Gao, N.D. Jin, W.X. Wang, Y.C. Lai, Phase characterization of experimental gas-liquid two-phase flows. Phys. Lett. A **374**, 4014–4017 (2010)
12. C.S. Daw, W.F. Lawkins, D.J. Downing, N.E. Clapp Jr, Chaotic characteristics of a complex gas-solids flow. Phys. Rev. A **41**(2), 1179–1181 (1990)
13. W.F. Lawkins, C.S. Daw, D.J. Downing, N.E. Clapp Jr, Role of low-pass filtering in the process of attractor reconstruction from experimental chaotic time series. Phys. Rev. E **47**(4), 2520–2535 (1993)
14. C.S. Daw, C.E.A. Finney, M. Vasudevan, N.A. van Goor, K. Nguyen, D.D. Bruns, E.J. Kostelich, C. Grebogi, E. Ott, J.A. Yorke, Self-organization and chaos in a fluidized bed. Phys. Rev. Lett. **75**(12), 2308–2311 (1995)

Chapter 9
Markov Transition Probability-Based Network for Characterizing Horizontal Gas-Liquid Two-Phase Flow

9.1 Methodology

More recently, we have employed a Markov transition probability-based network to characterize the flow behavior underlying horizontal gas-liquid two-phase flow [1]. We here introduce methodology and obtained results as follows: A Markov chain is a mathematical system that undergoes transitions from one state to another, between a finite or countable number of possible states. The Markov chain is a discrete-time process of memoryless feature in the sense that, given the past states and the present state, the future behavior (or the next state) depends only on the present state and not on the sequence of events that preceded it. Usually, the Markov chain can be described by a directed graph, in which the edges are labeled by the Markov probabilities of going from one state to the other states. According to the network generation method based on Markov transition probability proposed in references [2, 3], we can create the network representations of a time series as follows: Given a time series $x(t)$, we identify its Q quantiles which divide the range of signal values into Q intervals with equal probability (q_i, $i = 1,2,..., Q$). Each interval is represented by a node in the generated complex network, correspondingly, the generated network has Q nodes. Nodes n_i and n_j are connected if the value of $x(t)$ in interval q_i changes to that in interval q_j in one time step. A weight w_{ij} of an edge is the transition probability in a Markov model estimated from the time series. In particular, the weight w_{ij} for the edge from node n_i to node n_j is determined by the probability that a point in interval q_i is followed by a point in interval q_j. Repeated transitions between intervals lead to edges in the generated network with larger weights. We in Fig. 9.1 demonstrate the construction of a directed weighted complex network from a time series. As can be seen, the time series are divided into 4 intervals, i.e., q_1, q_2, q_3, q_4, consequently, the equivalent network of the time series has four nodes. Since the value of $x(t)$ in interval q_1 changes to that in interval q_2 in one time step, node 1 and node 2 are connected in the generated network. The number of times for the transitions from interval q_1 to all other intervals is 4 and the number of times for the transitions from interval q_1 to interval q_2 is 3. Consequently, the weight w_{12} for the edge from node 1 to node

Z.-K. Gao et al., *Nonlinear Analysis of Gas-Water/Oil-Water*
Two-Phase Flow in Complex Networks, SpringerBriefs on Multiphase Flow,
DOI: 10.1007/978-3-642-38373-1_9, © The Author(s) 2014

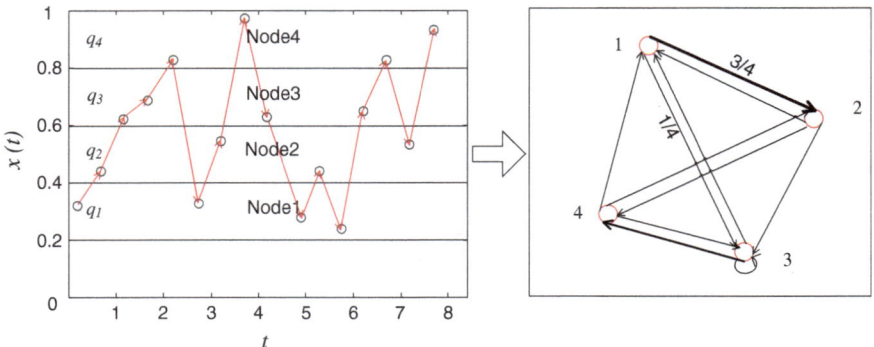

Fig. 9.1 The schematic diagram for the construction of a directed weighted complex network from a time series $x(t)$

2, i.e., transition probability from interval q_1 to interval q_2, is 3/4. Therefore, we can calculate the weights for all the edges of the generated complex network, i.e., obtaining the directed weighted network matrix. More details about the selection of Q quantiles refer to references [2, 3]. We using the above method generate networks from periodic sinusoidal time series, white noise time series, chaotic time series, respectively (as shown in Fig. 9.2). The chaotic time series are the data from the x component of Lorenz system

$$\begin{cases} dx/dt = \sigma(y - x) \\ dy/dt = x(\rho - z) - y \\ dz/dt = xy - \beta z \end{cases} \tag{9.1}$$

where $\sigma = 16, \rho = 45.92, \beta = 4$ as shown in Fig. 9.2c, and the data from the x component of Rössler system

$$\begin{cases} dx/dt = -y - z \\ dy/dt = x + ay \\ dz/dt = b + z(x - c) \end{cases} \tag{9.2}$$

where $a = 0.2, b = 0.2, c = 5.7$ as shown in Fig. 9.2d. We in Fig. 9.3 show the topological structures of the directed weighted networks inferred from four time series with different dynamics. It should be mentioned here that the network structures in Fig. 9.3 are drawn by the software "Ucinet" and "Netdraw", and the algorithm used is the Kamada-Kawai spring embedding algorithm [4].

As can be seen in Fig. 9.3a and b, the generated network inherits main properties of the time series in the network structure and the networks from time series with different dynamics exhibit distinct topological properties. In particular, the network inferred from the periodic time series exhibits regular attributes, as shown in Fig. 9.3a. In contrast, the network generated from the Gauss white noise time series presents random properties in the sense that the nodes in the network

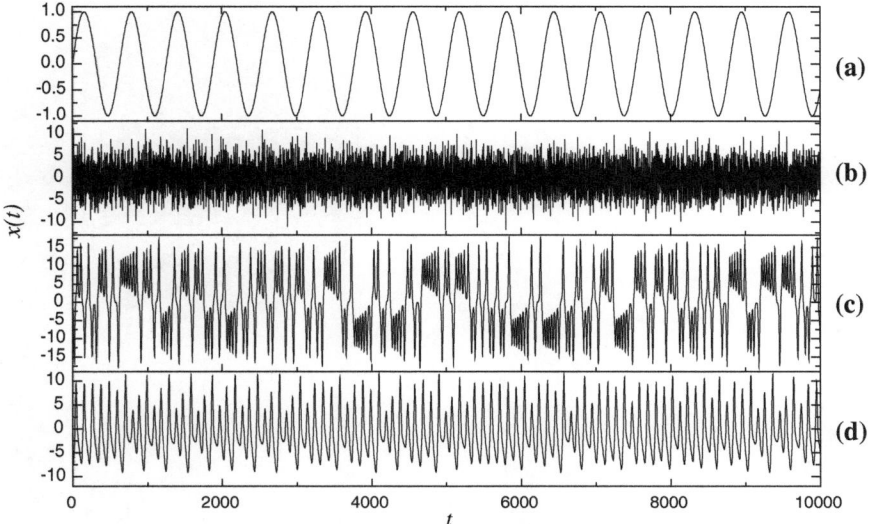

Fig. 9.2 Four time series with different dynamics for the construction of complex networks. **a** Periodic sinusoidal time series; **b** Gauss white noise time series; **c** Chaotic time series from Lorenz system; **d** Chaotic time series from Rössler system

are entangled with each other and the edges exist in an intersectional form, as can be seen in Fig. 9.3b. The complex networks from chaotic time series show the network configurations which are very similar to the chaotic attractor in phase space, i.e., the Lorenz and Rössler chaotic attractor, as shown in Fig. 9.3c and d. Specifically, the nodes in chaotic network congregate at different locations, leading to the existence of highly clustered regions and rather sparse regions. Distinguishing chaotic time series form random time series is a problem of continuous interest. Now we will demonstrate how to solve this problem in terms of topological statistics of the constructed network, i.e., the weighted clustering coefficient. The weighted clustering coefficient of a node i measures how densely connected the neighborhood of the node i is [2], and can be calculated as follows:

$$C_i = \frac{\sum_{j,k} w_{ij} w_{jk} w_{ki}}{\sum_{j,k} w_{ij} w_{ki}} \quad (9.3)$$

where the w_{ij} is the weight for the edge from node i to node j. The weighted network clustering coefficient is the average of weighted clustering coefficient for all nodes in the network, denoted as $<C>$. We infer networks from the time series of length 60000 from chaotic Lorenz system and white noisy system, respectively, and then calculate the weighted network clustering coefficient $<C>$ for the generated networks, as shown in Fig. 9.4. As can be been, due to the existence of unstable periodic orbits embedded in the chaotic attractor, the $<C>$ of chaotic network is obvious larger than that of noisy network, which allows distinguishing chaotic time series from noisy time series. Furthermore, we investigate the anti-

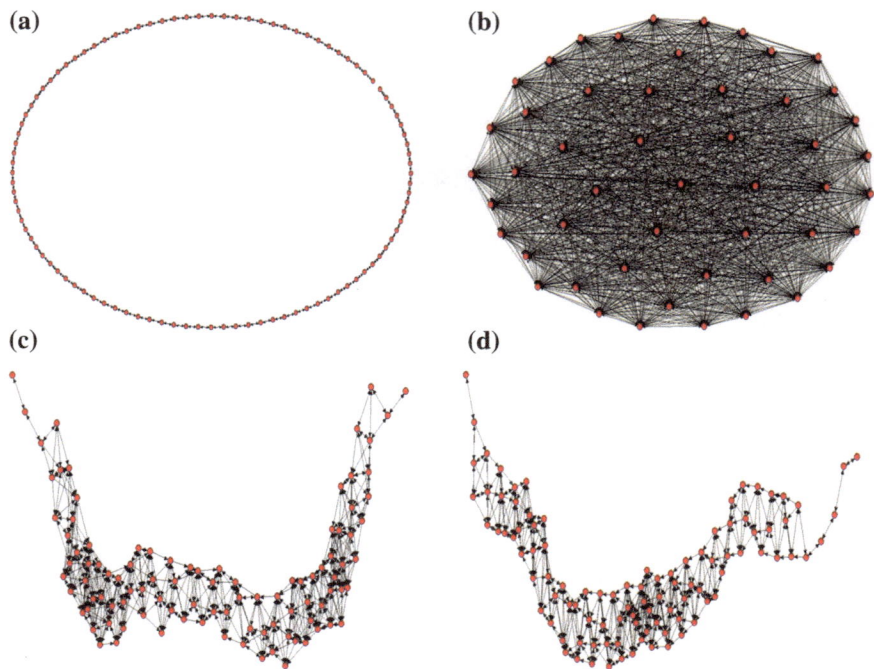

Fig. 9.3 The topological structures of the complex network generated from time series of length 10000 from **a** Periodic sinusoidal system; **b** Gauss white noise system; **c** Chaotic Lorenz system and **d** Chaotic Rössler system. The structural skeletons of various networks are drawn by the software "Ucinet" and "Netdraw"

Fig. 9.4 Distributions of the weighted clustering coefficient of the networks generated from various time series of length 60000

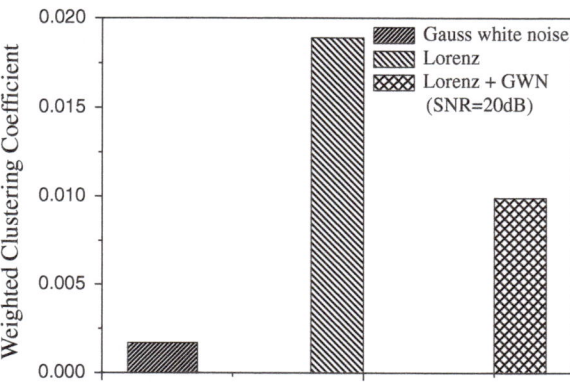

noise ability of the method by adding Gauss white noise to the chaotic Lorenz time series to generate a time series whose signal-to-noise ratios is 20 dB. We find that, despite the intense influence of noise, the weighted clustering coefficient $<C>$ is essentially larger than that of network from noisy time series, as shown in Fig. 9.4, indicating the robustness of the method. For more details see Ref. [5].

9.2 Dynamical Characterization of Horizontal Gas-Liquid Flow Patterns

We using the network generation method based on Markov transition probability construct complex networks from experimental signals of length 20000. The structural skeleton of the networks generated from the signals of three typical flow patterns are shown in Fig. 9.5. As can be seen, the generated networks corresponding to different flow patterns exhibit different topological properties. In order to characterize the dynamic behavior in the transition among different flow patterns, we construct numbers of networks from various experimental signals. We then calculate the weighted network clustering coefficient for each generated network. Figure 9.6 shows the distributions of weighted network clustering coefficient in semi-log scale for numbers of generated networks with the increase of gas superficial velocity, in which the U_{sg} and U_{sw} represent gas superficial velocity and water superficial velocity, respectively.

In order to further investigate the dynamical complexity in the transition among different flow patterns, we introduce the sample entropy [6] for quantifying their intrinsic characteristics. The sample entropy can be calculated by the following steps: (a) The time series $\{x(i), i = 1, 2, \cdots, n\}$, where n in length of data, can form a m-dimensional template vector $\overrightarrow{X_m^\tau}(i)$ by ordinal

$$\overrightarrow{X_m^\tau}(i) = \{x_m^\tau(i), x_m^\tau[(i+1)], \cdots, x_m^\tau[(i+m-1)]\}, i = 1 \sim n - m + 1 \qquad (9.4)$$

(b) Calculate the distance between the vector $\overrightarrow{X_m^\tau}(i)$ and $\overrightarrow{X_m^\tau}(j)$ for each i, the maximum difference of their corresponding scalar components can be defined as:

$$d\left[\overrightarrow{X_m^\tau}(i), \overrightarrow{X_m^\tau}(j)\right] = \max\{|x_m^\tau[(i+l)] - x_m^\tau[(j+l)]| : l = 0 \sim m - 1\} \qquad (9.5)$$

(c) Given a tolerance for accepting matches r, do the statistics of $B_i^m(r, n)$ for each i.

$$B_i^m(r, n) = \left\{\text{number of } j \text{ such that } d\left[\overrightarrow{X_m^\tau}(i), \overrightarrow{X_m^\tau}(j)\right] < r\right\} \bigg/ (n - m) \qquad (9.6)$$

For each $1 \leq i \leq n - m$, let $B_i^m(r, n)$ be the relative frequency to find a vector $\overrightarrow{X_m^\tau}(j)$ whose distance to $\overrightarrow{X_m^\tau}(i)$ is in a tolerance level r and $j \neq i$.

(d) The average of $B_i^m(r, n)$ can be calculated:

$$B^m(r, n) = (n - m)^{-1} \sum_{i=1}^{n-m} B_i^m(r, n) \qquad (9.7)$$

(e) Then we increase over one step from m to $m + 1$ and repeat the steps (b)–(d) to get $B^{m+1}(r, n)$.

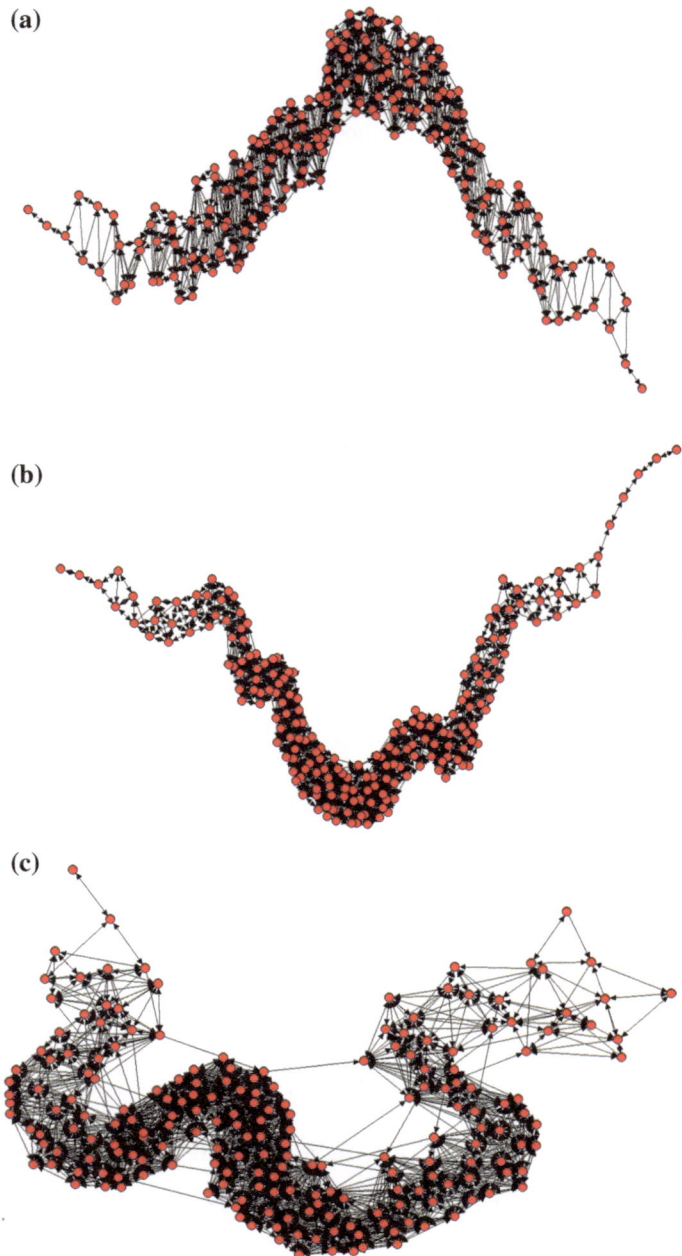

(a)

(b)

(c)

Fig. 9.5 The structural skeleton of the complex network generated from experimental signals of horizontal gas-liquid **a** Stratified flow; **b** Stratified wavy flow; **c** Slug flow

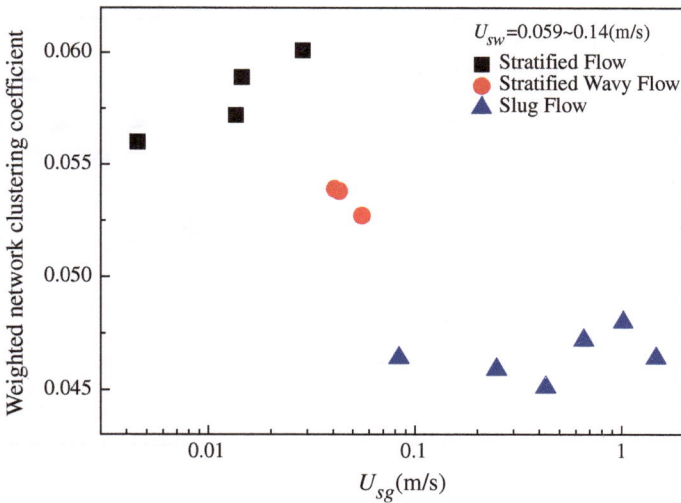

Fig. 9.6 The distributions of weighted network clustering coefficient for generated networks from different horizontal gas-liquid flow patterns

(f) Consequently sample entropy can be calculated:

$$SampEn(m, r, n) = \lim_{n \to \infty} \left\{ - \ln \left[B^{m+1}(r, n) / B^m(r, n) \right] \right\} \qquad (9.8)$$

Finally, in our analysis we use $m = 2$, $\tau = 1$ and $r = 0.2\sigma$ where σ is the standard deviation of the original time series. For a discussion on optimal choices for m and r, see the literature [7].

We show in Fig. 9.7 the distributions of sample entropy with the increase of gas superficial velocity. We find that there exists good relationship between weighted network clustering coefficient and sample entropy, i.e., the weighted network clustering coefficient is closely related to sample entropy, as shown in Figs. 9.6 and 9.7. When the gas superficial velocity is low, the flow structure of stratified flow is relative stable, the upper part of the pipe is gas phase and the bottom part of the pipe is the water phase. There does not exist any fluctuations in the phase interface, as shown in Fig. 2.2a and b. Correspondingly, the network clustering coefficients of stratified flow are large and the sample entropy is of small value, as shown in Figs. 9.6 and 9.7. With the increase of gas superficial velocity, the flow pattern evolves from stratified flow to stratified wavy flow. In this flow pattern transition, due to the increase of flow rate, the turbulence energy will increase. Consequently, the flow structure of stratified wavy flow become unstable gradually and the fluctuations appear in the interface between gas phase and water phase, as shown in Fig. 2.2c and d. Correspondingly, the network clustering coefficient decreases and the sample entropy increases as the flow pattern evolves from stratified flow to stratified wavy flow, as shown in Figs. 9.6 and 9.7. When the gas

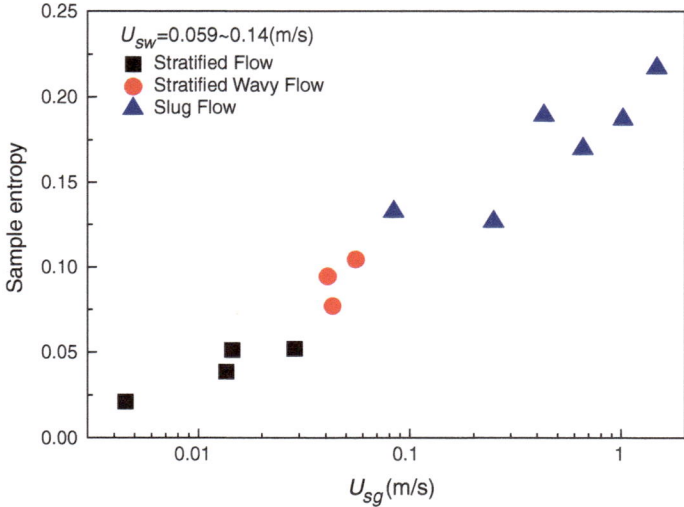

Fig. 9.7 The distributions of sample entropy for different horizontal gas-liquid flow patterns

superficial velocity is high, horizontal slug flow appears with the phenomenon of obvious fluctuations. Because of the influence of the turbulence effect, the flow structure of slug flow is very unstable, and compared with that of stratified wavy flow, the interface fluctuations of slug flow become strengthened, resulting from the existences of alternating movements between gas plug and liquid plug, as shown in Fig. 2.2e and f. Consequently, the network clustering coefficient will decrease and the sample entropy will increase in the transition from stratified wavy flow to slug flow, as shown in Figs. 9.6 and 9.7. Therefore, the network clustering coefficient is very sensitive to the flow structure associated with the interface fluctuations and allows investigating the dynamic flow behavior underlying different horizontal gas-liquid flow patterns.

References

1. L.D. Hu, N.D. Jin, Z. K. Gao, Characterization of horizontal gas-liquid two-phase flow using Markov model-based complex network. Int. J. Mod. Phys. C **24**(5), 1350028 (2013)
2. A.H. Shirazi, G.R. Jafari, J. Davoudi, J. Peinke, M.R.R. Tabar, M. Sahimi, Mapping stochastic processes onto complex networks. J. Stat. Mech.-Theory Exp., P07046 (2009)
3. A.S.L.O. Campanharo, M.I. Sirer, R.D. Malmgren, F.M. Ramos, L.A.N. Amaral, Duality between time series and networks. PLoS ONE **6**, e23378 (2011)
4. T. Kamada, S. Kawai, An algorithm for drawing general undirected graphs. Inform. Process. Lett. **31**(1), 7–15 (1989)
5. Z.K. Gao, L.D. Hu, N.D. Jin, Markov transition probability-based network from time series for characterizing experimental two-phase flow. Chin. Phys. B **22**(5), 050507 (2013)

6. J.S. Richman, J.R. Moorman, Physiological time-series analysis using approximate entropy and sample entropy. Am. J. Physiol. Heart Circ. Physiol. **278**(6), 2039–2049 (2000)
7. D.E. Lake, J.S. Richman, M.P. Griffin, J.R. Moorman, Sample entropy analysis of neonatal heart rate variability. Am. J. Physiol. Regul. Integr. Comp. Physiol. **283**(3), R789–R797 (2002)

Chapter 10
Recurrence Network for Characterizing Bubbly Oil-in-Water Flows

10.1 Recurrence Network Analysis of Time Series from Dynamic System

More recently, we have used the recurrence network to characterize the flow behavior of bubbly oil-in-water flows [1]. We here introduce methodology and obtained results as follows: Mapping a time series into a complex network allows quantitatively characterizing the structural characteristics of complex systems that are composed of a large numbers of entities interacting with each other in a complex manner. Now we introduce how to transform a time series into a complex network in the framework of recurrence network. When dealing with a time series $x(t)$ $(t = 1,...,M)$, we can use a proper m-embedding dimension and a suitable τ-time delay to reconstruct $x(t)$ [2],

$$\vec{X}(t) = (x(t), x(t+\tau), \cdots\cdots, x(t+(m-1)\tau)) \tag{10.1}$$

to obtain a phase space with N phase space vector points $\vec{X}(t), t = 1, 2, ..., N$, where

$$N = M - (m-1)\tau \tag{10.2}$$

Specifically, we use the method based on distinguishing false nearest neighbors (FNNs) [3] to determine the embedding dimension m and employ the correlation-integral-based algorithm [4] to determine the delay time τ. According to Refs. [5–7], in the framework of recurrence network, a phase space vector $\vec{X}(t_i)$ is said to be recurrent if there is $t_j \neq t_i$ such that

$$d\left(\vec{X}(t_i), \vec{X}(t_j)\right) < \varepsilon \tag{10.3}$$

where $d(\cdot, \cdot)$ is the distance measure in phase space

Z.-K. Gao et al., *Nonlinear Analysis of Gas-Water/Oil-Water*
Two-Phase Flow in Complex Networks, SpringerBriefs on Multiphase Flow,
DOI: 10.1007/978-3-642-38373-1_10, © The Author(s) 2014

$$d\left(\vec{X}(t_i), \vec{X}(t_j)\right) = \left\|\vec{X}(t_i) - \vec{X}(t_j)\right\| \qquad (10.4)$$

Then we can obtain the recurrence matrix that represents the configuration of recurrences in the phase space in terms of $d(\cdot, \cdot)$ as follows

$$R_{ij} = \Theta\left(\varepsilon - d\left(\vec{X}(t_i) - \vec{X}(t_j)\right)\right) \qquad (10.5)$$

where $\Theta(\cdot)$ is the Heaviside function. Consequently, we can infer an un-weighted and un-directed complex network from a given time series $x(t)$ by interpreting the recurrence matrix as the network adjacency matrix. In particular, in order to avoid self-loops in the network, the adjacency matrix of the recurrence network can be defined as:

$$A_{ij} = R_{ij} - \delta_{ij} \qquad (10.6)$$

where δ_{ij} is the Kronecker delta introduced in [5–7]. For a recurrence network, individual phase space vector serves as a node and the existence of an edge indicates the occurrence of a recurrence, i.e., the distance measure between a pair of nodes in the phase space is smaller than the threshold value ε. We using the above method generate a recurrence network from a chaotic time series from the x component of Duffing system

$$\ddot{x} + \delta\dot{x} - \beta x + \alpha x^3 = \gamma \cos wt \qquad (10.7)$$

where $\alpha = \beta = 1.0, \delta = 0.2, \gamma = 0.36$ and $w = 1.0$. We show the structure of the generated network in Fig. 10.1a. It should be noted that the network structure shown in Fig. 10.1a is drawn by the software "Pajek", and the computational algorithm is the Kamada-Kawai spring embedding algorithm [8]. The basic idea of the Kamada-Kawai spring embedding algorithm is as follows: First, the algorithm calculates the energy for every node in the network to find the one with the highest energy. Then, the algorithm iterates the Newton-Raphson stage to calculate and obtain new positions for the node until its energy is below epsilon. At this point, the algorithm again searches for the node of the highest energy and begins moving it. This process continues until the energy of every node is below epsilon, and then the algorithm is completed. For more details see Ref. [8]. As can be seen in Fig. 10.1a, the inferred recurrence network inherits the main features of the time series in its network configuration. In particular, the recurrence network from time series of chaotic Duffing system shows the network configuration which is very similar to the Duffing chaotic attractor in phase space, as shown in Fig. 10.1a and b.

As can be seen in Fig. 10.1a, the nodes in the chaotic recurrence network congregate at different locations resulting in the presence of some highly clustered regions and some sparse regions. A question arising here is how to associate the topological property of recurrence network with the dynamical characteristics of chaotic time series. Actually, the recurrence network can conserve many local features of the invariant density of the dynamic system captured in time series

(a)

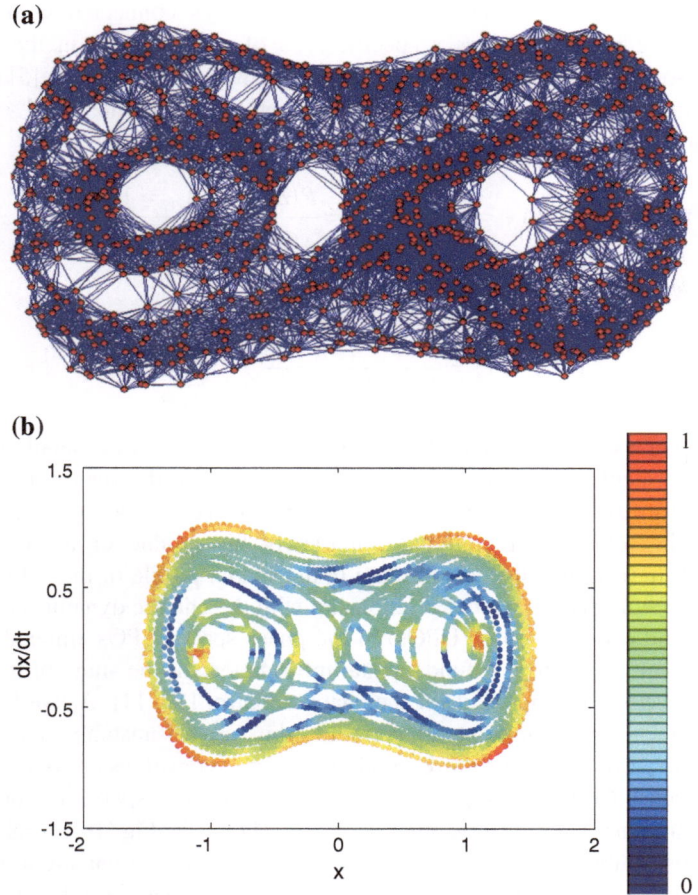

(b)

Fig. 10.1 Recurrence network generated from a time series of chaotic Duffing system. **a** The network structure drawn by Kamada-Kawai spring embedding algorithm; **b** *color-coded* representation of the local clustering coefficients $C^L(v)$ for the chaotic Duffing system

signals and, correspondingly, the local topological statistic of recurrence network allows characterizing the dynamics underlying time series. In this letter we use the local clustering coefficient defined by Donner et al. [5] to characterize the chaotic dynamics of time series. The clustering coefficient of a node v, $C(v)$, originally proposed by Watts and Strogatz [9], measures how densely connected the neighborhood of the node v is, and can be calculated as follows:

$$C(v) = \frac{2E_v}{k_v(k_v - 1)} = \frac{\sum_{j,m} A_{ij}A_{im}A_{mj}}{k_v(k_v - 1)} \qquad (10.8)$$

where E_v is the total number of closed triangles including node v, which is bound by the maximum possible value of $k_v(k_v - 1)/2$, where k_v is the degree of the node

v and the degree of a node is the number of edges connected with it. High clustering coefficient indicates a specific type of configuration in the network, which is closely related to the cliquish feature of a node. Donner et al. [5] extended this definition in terms of conditional probabilities to define local clustering coefficient, denoted as $C^L(v)$, as follows

$$C^L(v) = P(A_{ij} = 1 | A_{vi} = 1, A_{vj} = 1) = \frac{P(A_{ij} = 1, A_{vi} = 1, A_{vj} = 1)}{P(A_{vi} = 1, A_{vj} = 1)} \quad (10.9)$$

using Bayes' theorem, with

$$P(A_{vi} = 1, A_{vj} = 1) = \frac{1}{(N-1)(N-2)} \sum_{i=1}^{N} \sum_{j=1, j \neq i}^{N} A_{vi} A_{vj} \quad (10.10)$$

and a similar expression for $P(A_{ij} = 1, A_{vi} = 1, A_{vj} = 1)$. More details see Refs. [5–7]. We calculate the local clustering coefficient for the recurrence network shown in Fig. 10.1a and present the result in Fig. 10.1b. As can be seen, the node in the highly clustered region has remarkably high value of local clustering coefficient. In fact, this distinct structure in the spatial profile of the network local clustering coefficient has a close relationship with the chaotic dynamics associated with unstable periodic orbits (UPOs) in the phase space. UPOs embedded in the chaotic attractor can provide fundamental information for the study of the chaotic dynamics, especially important for the chaotic control [10, 11]. As we know, in a chaotic attractor the infinite set of periodic orbits are all unstable, and their stabilities as determined by the corresponding largest eigenvalues are typically quite heterogeneous. Correspondingly, we can see dense and sparse regions in the network structure from chaotic time series, as shown in Fig. 10.1a, and heterogeneous distribution of the local clustering coefficient in the chaotic attractor, as shown in Fig. 10.1b. In fact, a set of UPOs are essentially less unstable than others. A chaotic trajectory in the phase space will tend to spend more time around weak unstable orbits. As a result, there exist many recurrences near such a UPO, giving rise to a cluster in the corresponding network, which can be effectively captured by a high value of local clustering coefficient in the sense that the neighborhoods of a node in this region usually densely connected with each other. Donner et al. [5–7] have demonstrated that the regions of increased local clustering coefficient usually coincide with low-periods UPOs. In addition, high-periods UPOs also could be detected in terms of local clustering coefficient in the limit of $\varepsilon \to 0$ and $N \to \infty$. In this regard, the local clustering coefficient of recurrence network can be a faithful tool that allows characterizing chaotic dynamics associated with low-order unstable periodic orbits from time series.

10.2 Dynamic Characterization of Flow Patterns

We construct oil-in-water flow patterns recurrence networks from experimental signals. Transforming a time series into a complex network, we can investigate the inherent structure and dynamical characteristics of a time series in terms of the knowledge from complex network theory. In particular, representing an experimental signal through a corresponding recurrence network, we can then study the dynamical behaviors of vertical upward oil-in-water flow patterns from complex network analysis, which is quantified via local clustering coefficient. Figure 10.2 shows the structures in the spatial profile of the local clustering coefficient for the recurrence networks generated from the signals of three typical oil-in-water flow patterns, i.e., oil-in-water slug flow, oil-in-water bubble flow and very fine dispersed oil-in-water bubble flow. From Fig. 10.2, we can extract the mean value of local clustering coefficients for the network, denoted as $<C^L(v)>$, which can be calculated by

$$<C^L(v)> = \frac{1}{N}\sum_{v=1}^{N} C^L(v) \tag{10.11}$$

where N is the number of nodes in the network. A signal measured from one flow condition corresponds to a recurrence network. Note that the flow condition here refers to the flow behavior under different ratios between oil flow rate and water flow rate in the pipe. One flow condition corresponds to one certain flow pattern but one flow pattern may include different flow conditions. In order to characterize the dynamic flow behavior in the transition of different flow patterns, we construct numbers of recurrence networks from various experimental signals measured from different flow conditions. We then calculate the mean value of local clustering coefficients for each generated network. Figure 10.3 displays the distributions of $<C^L(v)>$ for numbers of generated networks with the change of total mixture velocity under five different water flow rate fraction, in which the K_w and U_{total} represent water flow rate fraction and total velocity of the oil-water mixture flow, respectively. We could see that, the $<C^L(v)>$ of oil-in-water slug flow are usually of large values, and $<C^L(v)>$ will decrease in the transition from oil-in-water slug flow to oil-in-water bubble flow. With a further increase in the total velocity of the oil-water mixture flow, the $<C^L(v)>$ will further decrease as the flow pattern evolves from oil-in-water bubble flow to very fine dispersed oil-in-water bubble flow. In addition, the $<C^L(v)>$ for oil-in-water slug flow tend to increase as the water flow rate fraction increasing from 80 to 92 %.

We now demonstrate how to quantitatively characterize the dynamic flow behavior underlying different vertical upward oil-in-water flow patterns in terms of network local clustering coefficients shown in Figs. 10.2 and 10.3. Donner et al. [5–7] have indicated that the local clustering coefficient in recurrence networks is closely related to a notion of local dimensionality of the point set generated by the dynamical system's trajectory in phase space [6]. The formula relating local

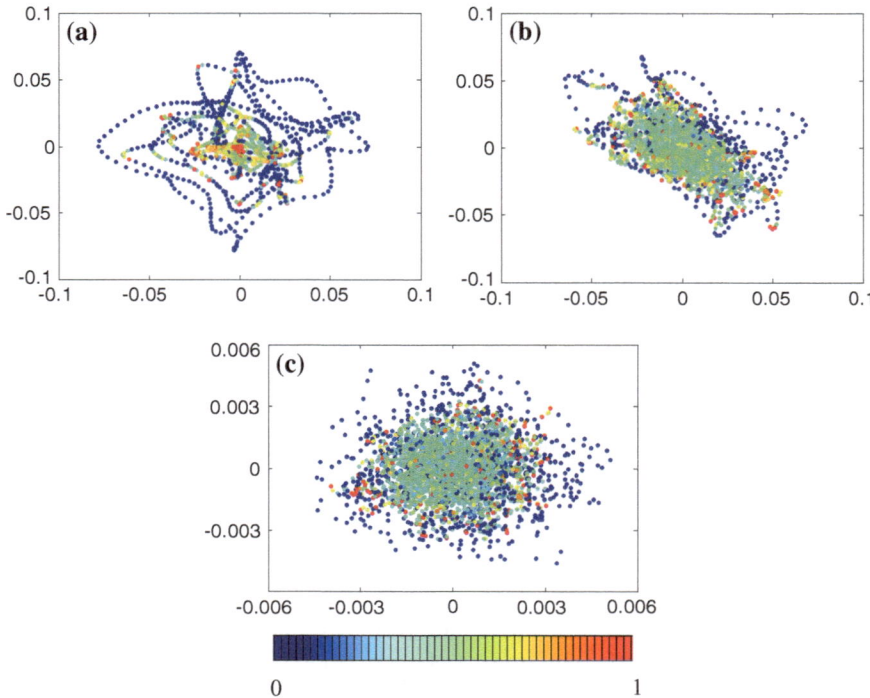

Fig. 10.2 Color-coded representation of the local clustering coefficients $C^L(v)$ in the phase space for the recurrence network generated from three types of oil-in-water flow patterns. **a** Oil-in-water slug flow ($K_W = 0.88$, $U_{total} = 0.01842$ m/s); **b** oil-in-water bubble flow ($K_W = 0.88$, $U_{total} = 0.07368$ m/s); **c** very fine dispersed oil-in-water bubble flow ($K_W = 0.88$, $U_{total} = 0.18421$ m/s)

clustering coefficient C and local clustering dimension D is $C = 0.75^D$, when using the supremum norm in phase space to define recurrences. In this regard, when the local dimensionality is low everywhere, the mean value of local clustering coefficient $<C^L(v)>$ is large and vice versa [6]. In particular, a quasi-periodic dynamics leads to a low dimensional attracting set in phase space, and, hence, to a large clustering coefficient. Vertical upward oil-in-water slug flow occurs at low oil-water mixture flow rate, where the small oil bubbles in water continuous phase coalesce to form oil slugs in different sizes. With the coalescence of small oil bubbles becomes more and more, a large number of oil slugs appear. Because of the quasi-periodic oscillation behavior associated with chaotic dynamics between oil slugs, the distributions of local clustering coefficients of oil-in-water slug flow present strong heterogeneous, as shown in Fig. 10.2a, correspondingly, the $<C^L(v)>$ of oil-in-water slug flow are of large values (i.e., low local dimensionality in the phase space), as shown in Fig. 10.3, which well agrees with the results presented in Chap. 3 and the discussions in Ref. [6]. With the increase of total velocity of the oil-water mixture flow (U_{total}), the turbulent kinetic energy is

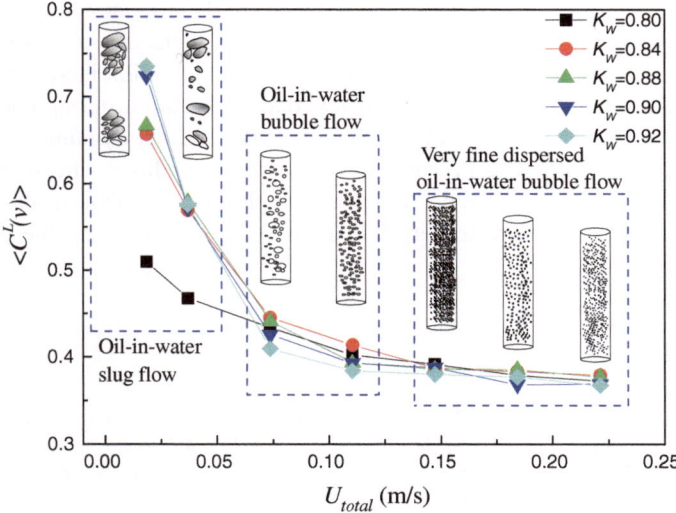

Fig. 10.3 Distributions of $<C^L(v)>$ for recurrence networks generated from different flow conditions

strengthened and the oil slugs are dispersed into small oil bubbles, i.e., oil-in-water bubble flow occurs. In this flow pattern, the oil phase exists in the form of discrete bubbles in water continuous phase. Owing to the stochastic motions of large numbers of small oil bubbles, the dynamical behavior of oil-in-water bubble flow become more random and complex than that of oil-in-water slug flow and no obvious periodic orbits exist in the phase space from oil-in-water bubble flow, corresponding to the weak heterogeneous distributions of local clustering coefficients, as shown in Fig. 10.2b, and small value of $<C^L(v)>$, as shown in Fig. 10.3. When the U_{total} is high, very fine dispersed oil-in-water bubble flow gradually occurs. Due to the influence of high turbulent kinetic energy, the oil bubbles are broken into smaller oil droplets in the transition from oil-in-water bubble flow to very fine dispersed oil-in-water bubble flow. The motions of large numbers of smaller oil droplets become rather stochastic, and correspondingly the dynamical behavior of very fine dispersed oil-in-water bubble flow becomes more complex than that of oil-in-water bubble flow. Consequently, the value of $<C^L(v)>$ decreases and the heterogeneity of the distributions of local clustering coefficients becomes weak as the flow pattern evolves from oil-in-water bubble flow to very fine dispersed oil-in-water bubble flow, as shown in Figs. 10.2b, c and 10.3. In addition, we can see that, with the increase of U_{total}, oil bubbles in the mixture flow become smaller and smaller in the evolution of three types of oil-in-water flow patterns, correspondingly, the $<C^L(v)>$ gradually decreases as the oil bubble becomes smaller, as shown in Fig. 10.3, indicating that the local clustering coefficients of recurrence network is very sensitive to the oil bubble size in the transition of flow patterns. Furthermore, it should be pointed out that, with the increase of water flow rate fraction (K_w), the global determinacy of the motions of oil slugs tend to be gradually strengthened, which explains the $<C^L(v)>$ for oil-in-water slug flow tend to increase as the water flow rate fraction increasing from 80 to 92 %. These interesting findings suggest that the local clustering coefficient of the recurrence network can faithfully represent the distinct dynamic states of the vertical upward bubbly oil-in-water flows and further allows effectively uncovering the dynamic flow behavior associated with chaotic UPOs in the transition from oil-in-water slug flow to very fine dispersed oil-in-water bubble flow.

References

1. Z.K. Gao, X.W. Zhang, M. Du, N.D. Jin, Recurrence network analysis of experimental signals from bubbly oil-in-water flows. Phys. Lett. A **377**, 457–462 (2013)
2. F. Takens, Dynamical systems and turbulence. *Lecture Notes in Mathematics*, vol. 898 (Springer, New York, 1981), pp. 366–381
3. T. Sauer, J.A. Yorke, M. Casdagli, Embedology, J. Stat. Phys. **65**, 579–616 (1991)
4. H.S. Kim, R. Eykholt, J.D. Salas, Nonlinear dynamics, delay times, and embedding windows. Physica D **127**, 48–60 (1999)
5. R.V. Donner, Y. Zou, J.F. Donges, N. Marwan, J. Kurths, Recurrence networks-a novel paradigm for nonlinear time series analysis. New J. Phys. **12**, 033025 (2010)
6. R.V. Donner, J. Heitzig, J.F. Donges, Y. Zou, N. Marwan, J. Kurths, The geometry of chaotic dynamics—a complex network perspective. Eur. Phys. J. B **84**, 653–672 (2011)
7. R.V. Donner, M. Small, J.F. Donges, N. Marwan, Y. Zou, R. Xiang, J. Kurths, Recurrence-based time series analysis by means of complex network methods. Int. J. Bifurcat. Chaos **21**(4), 1019–1046 (2011)
8. T. Kamada, S. Kawai, An algorithm for drawing general undirected graphs. Inform. Process. Lett. **31**(1), 7–15 (1989)
9. D.J. Watts, S.H. Strogatz, Collective dynamics of 'small-world' networks. Nature **393**(6684), 440–442 (1998)
10. E. Ott, C. Grebogi, J.A. Yorke, Controlling chaos. Phys. Rev. Lett. **64**, 1196–1199 (1990)
11. G.T. Kubo, R.L. Viana, S.R. Lopes, C. Grebogi, Crisis-induced unstable dimension variability in a dynamical system. Phys. Lett. A **372**, 5569–5574 (2008)

Chapter 11
Conclusions

Despite tremendous knowledge about fluid flows, our understanding of nonlinear dynamics in two-phase flows is still quite limited. We have proposed a scheme based on complex network to investigate gas-water and oil-water two-phase flow and have obtained a series of fascinating results. The last decade has witnessed a tremendous growth in the study of complex networks, and network-based theories and methodologies have been increasingly applied to addressing fundamental problems in many disciplines. Two-phase flow as a class of complex nonlinear dynamical systems is far from being well understood. Introducing and developing the complex network-based approaches can yield quantitative insights into the dynamics of two-phase flows, in addition to traditional approaches, particularly with respect to the transition among flow patterns. We believe that complex network theory can shed new light on a variety of complex systems, ranging from natural and biological to social and engineering field.

Z.-K. Gao et al., *Nonlinear Analysis of Gas-Water/Oil-Water*
Two-Phase Flow in Complex Networks, SpringerBriefs on Multiphase Flow,
DOI: 10.1007/978-3-642-38373-1_11, © The Author(s) 2014